# 180
# Advances in Polymer Science

# Advances in Polymer Science

Recently Published and Forthcoming Volumes

# Interphases and Mesophases in Polymer Crystallization I

Volume Editor: Giuseppe Allegra

With contributions by
D. C. Bassett · D. S. M. De Silva · P. H. Geil · T.-C. Long · B. Lotz
K. L. Petersen · E. G. R. Putra · S. Rastogi · M. A. Shcherbina
A. E. Terry · G. Ungar · A. J. Waddon · R. A. Williams · P. Xu
J. Yang

 Springer

This series presents critical reviews of the present and future trends in polymer and biopolymer science including chemistry, physical chemistry, physics and material science. It is adressed to all scientists at universities and in industry who wish to keep abreast of advances in the topics covered.

As a rule, contributions are specially commissioned. The editors and publishers will, however, always be pleased to receive suggestions and supplementary information. Papers are accepted for "Advances in Polymer Science" in English.

In references Advances in Polymer Science is abbreviated *Adv Polym Sci* and is cited as a journal.

The electronic content of *Adv Polym Sci* may be found at springerlink.com

Library of Congress Control Number: 2005922054

ISSN 0065-3195
ISBN-10 3-540-25345-9 **Springer Berlin Heidelberg New York**
ISBN-13 978-3-540-25345-7 **Springer Berlin Heidelberg New York**
DOI 10.1007/b106919

**Springer is a part of Springer Science+Business Media**
springeronline.com
© Springer-Verlag Berlin Heidelberg 2005
Printed in Germany

The use of registered names, trademarks, etc. in this publication does not imply, even in the absence of a specific statement, that such names are exempt from the relevant protective laws and regulations and therefore free for general use.

Cover design: *Design & Production* GmbH, Heidelberg
Typesetting and Production: LE-TEX Jelonek, Schmidt & Vöckler GbR, Leipzig

Printed on acid-free paper   02/3141 YL – 5 4 3 2 1 0

## Advances in Polymer Science
## Also Available Electronically

For all customers who have a standing order to Advances in Polymer Science, we offer the electronic version via SpringerLink free of charge. Please contact your librarian who can receive a password or free access to the full articles by registering at:

springerlink.com

If you do not have a subscription, you can still view the tables of contents of the volumes and the abstract of each article by going to the SpringerLink Homepage, clicking on "Browse by Online Libraries", then "Chemical Sciences", and finally choose Advances in Polymer Science.

You will find information about the

– Editorial Board
– Aims and Scope
– Instructions for Authors
– Sample Contribution

at springeronline.com using the search function.

# Preface

Polymer crystallisation is a field of science whose widespread practical and technological implications add to its scientific relevance. Unlike most molecular substances, synthetic polymers consist of long, linear chains usually covering a broad distribution of molecular lengths. It is no surprise that only rarely may they give rise to regularly shaped crystals, if at all. As a rule, especially from the bulk state, polymers solidify as very tiny crystals interspersed in an amorphous matrix and randomly interconnected by disordered chains. How do these crystals form? Do they correspond to a state of thermodynamic equilibrium, or are the chains so inextricably entangled that equilibrium is virtually impossible to reach? There is currently a widespread consensus on the latter conclusion, which only makes the problem more interesting as well as more difficult to handle. The perspective at the base of the present endeavour can be summarised with two questions: What are the key structural steps from the original non-crystalline states to the semi-crystalline organization of the polymer? Do these different stages influence the resulting structure and to what degree?

As demonstrated by the collection of review articles published within three volumes of *Advances in Polymer Science* (Volumes 180, 181 and 191), this problem may be approached from very different sides, just as with the related topic of polymer melting, for that matter. Morphological and atomistic investigations are carried out through the several microscopic and scattering techniques currently available. X-ray, neutron and electron diffraction also provide information to unravel the structure puzzle down to the atomistic level. The same techniques also allow us to explore *kinetic* aspects. The fast development of molecular simulation approaches in the last few decades has given important answers to the many open problems relating to kinetics as well as morphology; in turn, statistical-mechanical studies try to make sense of the many experimental results and related simulations. In spite of several successes over 60 years or more, these studies are still far from providing a complete, unambiguous picture of the problems involved in polymer crystallisation. As one of the authors (an outstanding scientist as well as a very good friend) told me a couple of years ago when we started thinking about this project, we should not regard this book as the solution to our big problem – which it is not – but rather as a sort of "time capsule" left to cleverer and better-equipped scientists of generations to come, who will make polymer crystallisation completely clear.

Thanks to all the authors for making this book possible. Here I cannot help mentioning one of them in particular, Valdo Meille, who helped with planning, suggesting solutions and organising these volumes. Thank you, Valdo, your intelligent cooperation has been outstandingly useful.

Milan, February 2005                                                    Giuseppe Allegra

# Contents

# Contents of Volume 181

## Interphases and Mesophases in Polymer Crystallization II

Volume Editor: Giuseppe Allegra
ISBN: 3-540-25344-0

# Contents of Volume 191

## Interphases and Mesophases in Polymer Crystallization III

Volume Editor: Giuseppe Allegra
ISBN: 3-540-28280-7

Adv Polym Sci (2005) 180: 1–16
DOI 10.1007/b107230
© Springer-Verlag Berlin Heidelberg 2005
Published online: 29 June 2005

# On the Role of the Hexagonal Phase in the Crystallization of Polyethylene

D. C. Bassett

JJ Thomson Physical Laboratory, University of Reading, Whiteknights,
Reading RG6 6AF, UK
*d.c.bassett@rdg.ac.uk*

**Abstract** Polyethylene forms a two-dimensional hexagonal phase, stable at $\geq\sim$ 3 GPa depending on molecular length, which in recent years has been claimed to intervene in crystallization prior to the formation of the usual orthorhombic phase even at atmospheric pressure. This claim is evaluated and shown to be without substance. There is very little evidence that the theoretical possibility of thin lamellae being more stable in the hexagonal phase than the orthorhombic at atmospheric pressure, if the former has sufficiently low fold surface free energy, does occur in practice. But the existence of single crystals of the orthorhombic phase unambiguously shows that they did not have a hexagonal precursor; that would have made them threefold twins. The overwhelming mass of evidence is that orthorhombic and hexagonal phases crystallize independently in accordance with the phase diagram and kinetic competition during growth, as has been understood since the hexagonal phase was discovered.

**Keywords** Polyethylene · Crystallization · Hexagonal phase · Metastable phases · Size-related stability

# 1
# Introduction

The concept of precursors to polymeric crystallization other than evolving nuclei is one that has recurred regularly since the early days of the subject so far without substantiation. As far as pre-ordering in the melt is concerned, the plausible idea that it might contain regions of aligned molecules is not supported by detailed X-ray analysis [1]. In terms of developing crystallites, three recent proposals concern origins respectively, in spinodal decomposition [2], block precursors [3] and an intermediate metastable phase, such as the hexagonal of polyethylene, in crystallization from the melt [4]. While the last of these is the particular concern of this article, fundamental difficulties in accepting the first two hypotheses may also be pointed out. First, crystallization is, in almost all circumstances, heterogeneously nucleated whereas spinodal decomposition is a homogeneous mechanism, with its own striking and characteristic morphology, e.g. [5], quite distinct from that of crystallized polymers. Second, insofar as classical thermodynamics retains its relevance to atomic dimensions, there is no obvious free energy minimum offering exceptional stability to a small block as nucleation proceeds. A particular block size is merely one stage in the progressive reduction in free energy, once the critical nucleus has been exceeded, as more stems crystallize. The addition of each stem does represent a local free energy minimum, with the positive surface contributions increasingly offset, but there is no more significant minimum which would confer exceptional stability for a particular dimension of block. Nor does a possible mesomorphic structure, of lower free energy, offer additional stability to the embryo: if it did exist it would become the preferred mode of crystallization prevailing over the observed crystal structure. *A priori*, there is no reason here to expect that crystallization proceeds other than by progressive development of the critical nucleus. Nor, as discussed below, does hexagonal polyethylene, a claimed metastable precursor to the orthorhombic phase at atmospheric pressure [4], provide an exception to this scenario. Convincing evidence in favour of precursors other than conventional nuclei playing a role in polymeric crystallization has yet to be provided.

# 2
# Hexagonal Polyethylene

## 2.1
## Context

The remarkable lamellar morphology of polyethylene crystallized at high pressures $\sim 0.5$ GPa, with thicknesses in the micron range [6], and sometimes substantially higher [7], was eventually correlated with crystallization of a new phase of the polymer, first on thermodynamic evidence [8, 9] then confirmed by X-ray analysis [10]. The author has previously reviewed the work by which this was established and its wider context within polymeric crystallization [11]; this is still valid but is now supplemented by the important later discovery that lamellae of the hexagonal phase form circular [12]. Salient points are that two distinct crystallization processes were identified, at low and high pressures, following the recognition that the optical texture of polyethylene crystallized at high pressure differed from the spherulitic organization typical of growth at atmospheric pressure or *in vacuo* being spiky as in immature spherulites grown at low supercoolings (Fig. 1) [13, 14]. Moreover, these two different textures persisted, little changed, in products of crystallization at intermediate pressures, $\sim 0.3$ GPa, when one gave way to the other over a narrow temperature interval depending on molecular length [14]. The two textures were found to form, isobarically, in different, non-overlapping, ranges of supercooling, the hexagonal first, at lower values, and to have melting points differing by $\sim 8$ K according to their respective thin and thick constituent lamellae [14]. The two forms tended to occur separately, in adjacent areas, but when the orthorhombic phase did grow on an existing hexagonal lamella, it did so with sharply decreased thickness. Crucially, it was then shown that unlike crystallization at atmospheric pressure, which occurred in a single stage, that at high pressure occurred with two sequential exotherms and two associated volume changes [8, 9]. These thermodynamic data were consistent with there being two first order transitions when polyethylene crystallized at 0.5 GPa but only one at low pressures or *in vacuo*. The former circumstance corresponded, it was proposed, to sequential transformation first from the melt to a new 'intermediate' phase then from this to the orthorhombic form and a phase diagram, constructed from thermal data, published [8, 9].

In situ X-ray examination of crystallizing polyethylene, at high temperature and pressure, then confirmed this proposal in detail, showing that the wide-angle diffraction pattern changed abruptly with the optical texture [10]. That corresponding to the spherulitic texture was of the usual orthorhombic form while the new 'intermediate' phase had two-dimensional hexagonal symmetry, with an increased cross-sectional area per chain, but without

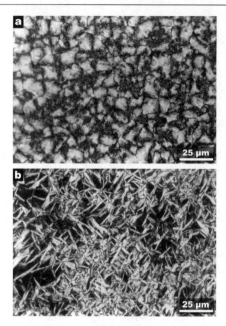

**Fig. 1** The differing optical textures, between crossed polars, of linear polyethylene after crystallization from the melt at pressures close to the triple point, $\sim 0.3$ GPa (**a**) the conventional spherulitic texture of the orthorhombic phase (**b**) the coarse lamellar texture formed as the hexagonal phase then transformed to orthorhombic during return to ambient temperature and pressure from [14]

a single chain configuration. Models in which both TTT and TGTG* configurations exist in the same chain, where T signifies trans, G and G* alternative gauche bond sequences, are able quantitatively to account for the experimental data, such as specific volume, of the hexagonal phase [15].

The distinct nature of the high and low pressure processes was subsequently reinforced with the demonstration that they give individual lamellae of different habits (Fig. 2). Orthorhombic polyethylene crystallizes from the melt with lamellae of familiar forms [16], showing some tendency to incipient dendritic growth, elongated along *a* and *b* axes, in the changeover region, and with molecules inclined at $\sim 35°$ to lamellae. Hexagonal polyethylene, in striking contrast, forms circular discs, thinner at their edges, to which molecules are normal [12]. These are so thick that they can be observed growing, individually, in the diamond-anvil pressure cell. The usual orientation presents lamellae in cross-section, with molecules approximately parallel to the diamond surfaces, consistent with flow during sample preparation. In this condition they display strong, clear birefringence contrast. The first observations reported that when a melt was subject to increased pressure lamellae 'flashed into view' [17] but when grown at low supercooling they can easily be held stable indefinitely. Only when the temperature is lowered sufficiently

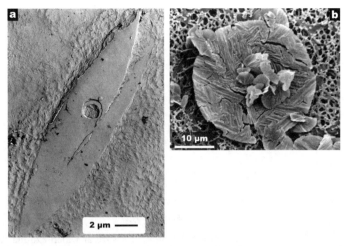

**Fig. 2** Single crystals of linear polyethylene (**a**) crystallized at 130 °C at atmospheric pressure (**b**) crystallized at 0.3 GPa as uniform circular discs then given a complex twinned texture with lines inclined at ∼ 60° during return to ambient temperature and pressure from [12]

does the birefringence change and the contrast become muddy when the hexagonal phase transforms to the orthorhombic and molecules incline to lamellae [18] leaving a characteristic record in the morphology with adjacent regions having their $b$ axes inclined at ∼ 60° [12]. This record is not found in lamellae of the orthorhombic form – as it would if they did have a hexagonal precursor – which are single crystalline.

## 2.2
## The Pattern of Crystallization

The phenomena described are in detailed accord with the phase diagram [9, 11] coupled with the concept of kinetic competition during growth. The phase diagram defines those regions in which a particular phase (of infinite size) is the most stable, having the lowest free energy (specific Gibbs function). Outside its boundary lines a given phase may still exist or form but in metastable condition. There is no requirement that only the stable crystalline phase can form within its region of the phase boundary. As always in crystal growth, it is the fastest growing path which is followed as, for example, in chainfolding and the phase which appears is not necessarily the most stable. In practice, when polyethylene crystallizes at high pressures for a typical cooling rate ∼ 1 K/min it is the hexagonal phase which forms first and metastably inside the orthorhombic-stable region (Fig. 3). However, circumstances may change this outcome which is not invariably the case: the orthorhombic phase forms directly from the melt on fast quenching [19]

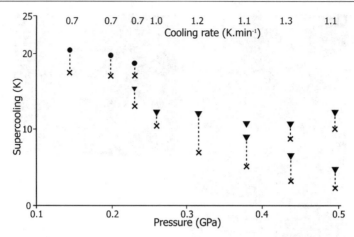

**Fig. 3** A plot of the supercoolings as a function of pressure at which exotherms appear during the crystallization of linear polyethylene during cooling from the melt at the rates shown. *Crosses* show the start of the exotherms; *filled circles* show the peak temperatures for orthorhombic crystallization; *filled triangles* show the sequential peak temperatures (where resolved) corresponding first to hexagonal crystallization then its conversion to the orthorhombic phase. Redrawn from [9]

while high molecular weight polyethylene cooled at 1 K/min and 0.5 GPa, crystallizes within the hexagonal-stable region [9].

Crystallization of the metastable phase is to be expected because the typical supercooling of $\sim 12$ K at which the hexagonal phase then forms (Fig. 3) is greater than the width of the hexagonal-stable region at 0.5 GPa [9]. However, when forming in the orthorhombic-stable region the hexagonal phase is inevitably in kinetic competition with the formation of that phase directly from the melt. At the mutual phase line this latter is the slower process but, with increasing supercooling the free energy falls more rapidly for the orthorhombic than for the hexagonal phase so that eventually direct crystallization of the orthorhombic phase will and does prevail [19].

Similar considerations apply to crystallization at pressures below the triple point [9, 11]. Here the melting point of the hexagonal phase (for infinite thickness) is lower than the orthorhombic so that the orthorhombic phase has the higher supercooling at a given temperature, increasingly so as the pressure falls further (Fig. 4). Experimental data show that, near the triple point $\sim 0.3$ GPa, with a cooling rate $\sim 1$ K/min, the hexagonal phase crystallizes at $\sim 12$ K of supercooling and the orthorhombic at $\sim 16$ K. It is to be expected, therefore, that the hexagonal phase will continue to form first at this cooling rate and pressures reducing below the triple point until the respective supercoolings of 12 K for the hexagonal and 16 K for the orthorhombic phase occur at the same temperature. This is consistent with experiment.

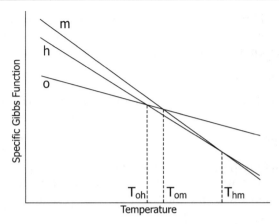

**Fig. 4** Schematic free energy diagram for the crystallization of polyethylene from the melt showing the specific Gibbs function (chemical potential) for melt (m), hexagonal (h), and orthorhombic (o), phases as function of temperature

The large body of explicit evidence cited shows that the hexagonal and orthorhombic phases crystallize in two distinct, independent processes. They form with different lamellar habits and optical textures, at distinct supercoolings for the same cooling rate, with the orthorhombic phase giving substantially thinner lamellae, melting $\sim 8$ K lower than those formed in the hexagonal phase which, moreover, commonly forms within the orthorhombic-stable region, entirely in accord with the phase diagram. Moreover, were the orthorhombic phase to form around a hexagonal precursor, the latter would leave a characteristic twinned morphology [12]; this has never been observed.

Nevertheless, it has subsequently been suggested that, for polyethylene, there could be a phase inversion at small lamellar thickness which could make the hexagonal the precursor of orthorhombic crystallization from the melt even at atmospheric pressure [4]. The basis of this proposal will now be outlined and the conclusion reached that it is inapplicable, in part because the effective fold surface energy during growth of hexagonal lamellae is not sufficiently low.

# 3
# Thickness-Related Stability

## 3.1
## Thermodynamics

The relative stabilities of orthorhombic and hexagonal phases are conveniently discussed using a free energy diagram as in Fig. 4. This plots specific Gibbs functions, $g$, (which are equal for infinite phases in equilibrium) against temperature, $T$, at constant pressure, $p$, for the two crystalline phases and the melt. From the fundamental relation

$$(\partial g/\partial T)_p = -s.$$

these are straight lines of slope $-s$, the specific entropy, if variations in specific heat capacities

$$c = T(\partial s/\partial T)$$

are ignored. Moreover, to cross a boundary in a phase diagram with increasing temperature at constant pressure, and achieve the necessary decrease of free energy requires that one moves to a phase of higher specific entropy, whence

$$s_m > s_h > s_o \tag{1}$$

referring to the specific entropies of melt, hexagonal and orthorhombic phases respectively and corresponding to the relative positions of the three phases in the phase diagram. This is reflected in the respective slopes of Fig. 4, in which the slope of the hexagonal line must lie between those of orthorhombic and melt.

Figure 4 shows relative free energies when the hexagonal phase is stable. This occurs when the orthorhombic and hexagonal lines intersect, at $T_{oh}$, below the orthorhombic melting temperature, $T_{om}$, followed by the melting of the hexagonal phase at $T_{hm}$. The interval $(T_{hm} - T_{om})$ is a measure of the relative stability of the hexagonal phase reflecting, as Fig. 4 shows, the difference in specific free energy of the two phases at $T_{om}$, and vanishing at the triple point, $T_t$, when all three lines have a common intersection. For lower temperatures, $T < T_t$, as mentioned above $T_{hm} < T_{om}$ and the hexagonal phase is metastable by an amount proportional to $(T_{om} - T_{hm})$.

## 3.2
### Stability Inversion with Lamellar Thickness

The lamellar habit adopted by crystalline polymers adds surface terms to the specific Gibbs function (chemical potential), most importantly the fold surface free energy, $\sigma_e$, which contributes $2\sigma_e/\lambda\varrho$ for a lamella of thickness $\lambda$ and crystalline density $\varrho$. In consequence melting points are lowered from $T_m^0$, for infinite thickness, to $T_m$ according to the Hoffman-Weeks equation

$$T_m = T_m^0(1 - 2\sigma_e/\lambda \cdot \Delta h_v) \tag{2}$$

where $\Delta h_v = \Delta h \cdot \varrho$ is the specific enthalpy of melting per unit volume of crystal (as opposed to $\Delta h$, the specific enthalpy per unit mass).

As a straightforward consequence, one may shift the lines in Fig. 4 upwards by $2\sigma_e/\lambda\varrho$ for each crystalline phase to obtain a modified diagram pertinent to lamellae of finite thickness $\lambda$. It then becomes possible, in principle, to move the orthorhombic line up sufficiently more than the hexagonal so as to 'invert' the stability and make a thin hexagonal lamella more stable than an orthorhombic one of the same thickness over a finite range of temperature surrounding the value of the orthorhombic melting point. Below this range, the orthorhombic is always the stable phase. Such stability inversion requires the surface contributions to free energy to be more favourable for the hexagonal than the orthorhombic phase.

If thin hexagonal lamellae are to be stable within the orthorhombic-stable region of the phase diagram, it is necessary, for the same considerations advanced above, that an orthorhombic lamella melts at a lower temperature than a hexagonal one of the same thickness, i.e. from Eq. 2

$$T_{mo} = T_{mo}^0(1 - 2\sigma_{eo}/\lambda \cdot \Delta h_{vo}) < T_{mh} = T_{mh}^0(1 - 2\sigma_{eh}/\lambda \cdot \Delta h_{vh}) \qquad (3)$$

where o and h suffices refer to orthorhombic and hexagonal phases respectively. Whence, the length below which the hexagonal phase is the more stable is

$$\lambda_s < 2\left[\sigma_{eo}T_{mo}^0/\Delta h_{vo} - \sigma_{eh}T_{mh}^0/\Delta h_{vh}\right]/(T_{mo}^0 - T_{mh}^0). \qquad (4)$$

The value of $\lambda_s$ tends to infinity as the denominator $(T_{mo}^0 - T_{mh}^0) \to 0$ as $T \to T_t$. Conversely, as $(T_{mo}^0 - T_{mh}^0)$ increases, i.e. the hexagonal phase becomes more metastable, $\lambda_s$ decreases as more surface contributions to the free energy, offset by the enthalpy of crystallization, are required, which according to the hypothesis must favour the hexagonal phase.

From Eq. 4, if $\lambda_s$ is to be positive,

$$\sigma_{eo}T_{mo}^0/\Delta h_{vo} > \sigma_{eh}T_{mh}^0/\Delta h_{vh}. \qquad (5)$$

### 3.3
### A Hexagonal Precursor?

The scheme proposed by Keller et al. [4] is that for melt crystallization of polyethylene in or near the range 122–130 °C at 1 bar lamellae form in the hexagonal phase, increase in thickness then transform to the orthorhombic after which there is little or no increase of thickness. Their discussion centres on Fig. 5, which plots melting point against inverse lamellar thickness for orthorhombic and hexagonal phases as well as their transition temperature[1]. For the ordinate, thermodynamics requires, as discussed above, that when the unrestricted hexagonal phase is unstable the melting point of infinitely thick lamellae is below that for the orthorhombic phase. If there is to be phase inversion, it is necessary that the lines in Fig. 5 must intersect at finite $\lambda$, at the

---

[1] We note a minor error by these authors who claim that Eq. 1 applies as written also to the orthorhombic/hexagonal transition. It does not but requires to be modified, as in Eq. 11, on account of the different densities of the two phases.

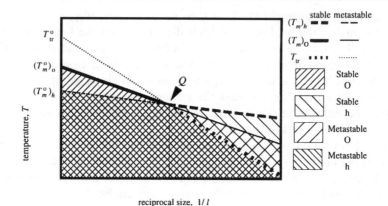

**Fig. 5** 'Phase-stability' diagram in which temperature is plotted against reciprocal thickness for polyethylene to illustrate the condition for phase inversion to appear with decreasing size. From [4]

point $Q$, the modified triple point. The geometrical condition for this to happen is simply that the negative slope of the orthorhombic line is more than that of the hexagonal, i.e.

$$T^0_{mo}(\sigma_e/\Delta h_v)_o > T^0_{mh}(\sigma_e/\Delta h_v)_h$$

which is Eq. 5 above[2].

There are two issues central to this proposal, namely: is there an inversion of phase stability at atmospheric pressure and does the hexagonal phase then crystallize before the orthorhombic? As will become clear, the available data do not allow a definite answer to the first but probably not, to the second the answer is certainly no. We consider these matters in turn, first testing the inequality 5 against measured parameters.

In Eq. 5, the values of $T^0_{mo}$ and $T^0_{mh}$ will differ by at most a few K so their ratio may safely be taken as unity in relation to the other terms involved, hence Eq. 5 effectively reduces to

$$\sigma_{eh}/\sigma_{eo} < \Delta h_{vh}/\Delta h_{vo} \qquad\qquad (6)$$

Because of the relative magnitudes of the specific enthalpies (see below) the implication is that

$$\sigma_{eh} \ll \sigma_{eo}$$

which has a certain *a priori* plausibility because $\sigma_e$ for polymer lamellae reflects not only lattice forces, as in side surface energies, but also the work of

---

[2] Keller et al. [4] mistakenly stated that the slope of the $T_m$ vs $1/\lambda$ plot from equation Eq. 2 is $(-2\sigma_e/\Delta h_v)$ omitting the factor $T^0_m$ and erroneously deduced the condition $(\sigma_e/\Delta h_v)_h < (\sigma_e/\Delta h_v)_o$ although this mistake has no significant effect on calculations.

chainfolding. Whereas this is high for orthorhombic polyethylene, which requires at least 3, and probably more, gauche bonds to be introduced into the chain, gauche bonds are already present in chains within the hexagonal structure so that the work of chainfolding and $\sigma_{eh}$ will be reduced accordingly. In practice, $\sigma_{eh}$ is certainly less than $\sigma_{eo}$ but it is not clear that the reduction is sufficient to give phase inversion under the operative conditions.

The ratio of the specific enthalpies of fusion may be estimated starting from the specific entropies which are

$$\Delta s_h = s_m - s_h \quad \text{and} \quad \Delta s_o = s_m - s_o .$$

From Eq. 1 we have

$$s_m > s_h > s_o$$

hence

$$\Delta s_h < \Delta s_o$$

and, therefore,

$$\Delta h_h < \Delta h_o \tag{7}$$

recalling that

$$\Delta h = T_m^0 \Delta s$$

at a transition.

Equation 7 refers to $\Delta h$, the specific enthalpy of melting which differs from $\Delta h_v$, the equivalent quantity per unit volume, by the factor $\varrho$, the crystalline density, a difference amounting to some 8% between orthorhombic and hexagonal phases [10].

Inserting $\Delta h_v = \varrho \Delta h$ into Eq. 7 gives

$$[\Delta h_v/\varrho]_h < [\Delta h_v/\varrho]_o$$

and

$$1.08 \Delta h_{vh} < \Delta h_{vo} \tag{8}$$

Combination of Eq. 6 and Eq. 8 leads to

$$\sigma_{eh}/\sigma_{eo} < \Delta h_{vh}/\Delta h_{vo} < 0.92 \tag{9}$$

provided that chains are normal to lamellae in both phases; if not $\lambda$ needs to be reduced by $\sim \cos 35° = 0.82$ for the orthorhombic phase.

While the right hand inequality of Eq. 9 certainly holds for polyethylene, the left hand one may not. The first, somewhat imprecise, measurements [9] estimated $\Delta h_h/\Delta h_o$ as $\sim 0.25$ whence $\Delta h_{vh}/\Delta h_{vo} \sim 0.23$ whereas the same paper estimated $\sigma_{eh}/\sigma_{eo} = 0.2 \pm 0.13$ from melting data.

On this basis and also using more recent values of the parameters [20], size inversion appears a marginal possibility – although this does not imply that it will be possible to crystallize lamellae of this thickness from the melt - but is not immediately ruled out notwithstanding the absence of positive evidence for its occurrence. However, in both instances, the values of $\sigma_e$ used were derived from melting data whereas, as was pointed out when the hexagonal phase was discovered [9], they should be for growth, when values will be higher. Recent work has revealed that this is a very significant factor, more so than had previously been thought.

It has been shown that, at atmospheric pressure, the fold surfaces of melt-crystallized polyethylene generally form rough then reorganize as fold packing seeks to move towards the preferred $\{201\}$ surface [21]. Only at the lowest supercoolings ($\geq 127\,^\circ$C for $\sim 10^5$ mass polymer at 1 bar and slow growth rates, $<\sim 1\,\mu$m/min) are folds able to attain a $\{201\}$ surface before the next molecular layer is added. When surfaces form rough, as they mostly do, it is because the energetic cost of poor fold packing is less of a consideration for a molecule adding to a lamella than the gains from crystallizing stems. The effective surface energy will then rise because of the rough geometry which, it is likely, will be the major factor reflected in the operative values of $\sigma_e$ in contrast to those pertinent to well-defined fold surfaces. In other words, the effective surface free energies for the two rival phases to grow with rough surfaces can confidently be expected to be closer in value than those measured during melting, which mitigates strongly against Eq. 9 being satisfied for rapid crystallization.

This situation would change for slow crystallization at atmospheric pressure and at high pressures near the triple point when orthorhombic lamellae form with $\{201\}$ surfaces and correspondingly lower surface energies. While this will improve the chances of Eq. 9 being satisfied and the change in effective surface energies could give an associated discontinuity in initial lamellar thickness vs crystallization temperature at 1 bar, this is not apparent in data recorded after isothermal thickening. Near the triple point, when lamellae of the two phases form at the same temperature and pressure, there is a considerable disparity in thickness: hexagonal lamellae are $\geq 100$ nm in thickness with orthorhombic ones melting 8 K lower corresponding to $\sim 30$ nm indicative of distinct processes of growth of the separate phases.

In general, the thickness of lamellae crystallized at supercooling $\Delta T$ is

$$\lambda = 2(\sigma_e/\Delta h_v)(T_m^0/\Delta T) + \delta\lambda. \tag{10}$$

The first term represents the condition of stability when the cost of surface terms, $2\sigma_e$, is just offset by the reduction in free energy from the volume term, $\Delta h_v$, while the second, $\delta\lambda$, relates to the need to maximize growth flux and crystallize at the fastest rate. It follows that, barring some extraordinary change in $\delta\lambda$ between the two phases, the greater thickness of hexagonal lamellae signifies a greater value of $(\sigma_e/\Delta h_v)$ for the hexagonal phase. Nor is the argument sustainable that the observed thickness of hexagonal lamellae is the result of substantial thickening from a dimension thinner than orthorhombic lamellae. This conflicts, *inter alia*, with observations on the melting and transitions of thin polyethylene lamellae.

# 4
# Transitions of Thin Polyethylene Lamellae

## 4.1
## Melting

A simple test of whether stability does invert with thickness which correctly uses melting-derived surface energies is directly to melt thin orthorhombic lamellae (10–15 nm, grown from solution) without thickening which is possible by judicious choice of heating rate. For melting, unlike crystallization, transitions are generally recorded at or close to the phase lines; there is little or no superheating because of the presence of suitable nuclei whereas considerable supercooling is the norm for crystallization. Were the hexagonal phase stable it would be revealed as a double transition in the melting endotherm around the value of the orthorhombic melting point, first from orthorhombic, then to the melt. The common experience is that this does not happen. When the hexagonal phase does appear at atmospheric pressure, for example, in highly oriented fibres [22] or after irradiation [23] its stability arises from a different cause namely reduced entropy of the melt. Nevertheless, Keller et al. [4] claim that there is X-ray evidence for the hexagonal phase appearing just prior to melting 15 nm thick lamellae at 1 bar; their detailed findings have still to be published. Even if correct, however, such stability would not of itself imply that hexagonal lamellae are able to satisfy Eq. 10 and form from the melt so thin.

## 4.2
## Annealing at High Pressure

Among the early experiments following the identification of the high pressure phase were studies of thin solution-grown polyethylene lamellae annealed at 0.5 GPa, adjacent to bulk specimens of the same polymer [24]. Although undertaken for different purposes, these bear significantly on the proposed role of the hexagonal phase in crystallization. They show, *inter alia*, a) that 15 nm thick lamellae enter the hexagonal phase at a depressed transition temperature with no evident thickening, b) that these lamellae melt, with only moderate thickening similar to that found for annealing in the orthorhombic phase, from the hexagonal phase with slight depression of melting point and c) that only on recrystallization from the melt into the hexagonal phase, is there a substantial increase of thickness, to $\sim 50$ nm

The first point, a), signifies that $\sigma_{eo}/\varrho_o > \sigma_{eh}/\varrho_h$, confirming that $\sigma_{eh} <$ $0.92\sigma_{eo}$; were the two surface contributions equal there would be no depression. The second point, b) shows that lamellae do not necessarily thicken substantially when in the hexagonal phase and that $\sigma_{eh}$ is small. By contrast, Keller et al. [4] assume that substantial thickening is inherent to the hexagonal phase. Though this is often asserted, with the the implication that this differs from the orthorhombic phase, the claim needs to be justified. It needs to be recalled that when both phases crystallize together near the triple point, the ratio of their respective lamellar thicknesses is about that of their enthalpies of crystallization and does not require exceptional thickening in the hexagonal phase. The further increases in lamellar thickness with increasing crystallization pressure could reflect increased molecular vibrations with pressure (and be present also in the orthorhombic phase at these pressures) rather than requiring exceptional thickening to be an inherent property of the hexagonal phase at all pressures. At present such matters are speculative but they will be pertinent to future theories of lamellar thickness which incorporate isothermal thickening and are not limited essentially to the size of the secondary nucleus. From the third point, c), it appears that $\sim 50$ nm is the lowest thickness at which the hexagonal phase forms directly from the melt, i.e. that this is the thickness at which the volume terms sufficiently offset the free energy penalty of the folded surfaces.

The evidence that lamellae 15 nm thick can enter the hexagonal phase came from electron diffraction of specimens annealed as low as 236 °C at 0.54 GPa, $\sim 3$ K lower than the maximum of the corresponding differential thermal analysis peak recorded for bulk polymer under the same conditions. At this annealing temperature lamellae were still entire, i.e. essentially of unaltered thickness although with some internal variation. No attempt was then made to anneal at lower temperatures and determine the orthorhombic/hexagonal transition temperature for this thickness. Such experiments offer, nevertheless, a direct approach by which the quantities of Eq. 9 may be determined.

The depression of the orthorhombic/hexagonal transition temperature for lamellae thickness $\lambda$ follows from equating the differential surface and compensating volume contributions to the free energy according to the geometry of Fig. 4. Thus

$$(2/\lambda)(\sigma_{eo}/\varrho_o - \sigma_{eh}/\varrho_h) = (T_{oh}^0 - T_{oh})(s_h - s_o)$$

i.e.

$$(T_{oh}^0 - T_{oh}) = [2/\lambda(s_h - s_o)] (\sigma_{eo}/\varrho_o - \sigma_{eh}/\varrho_h). \tag{11}$$

Equation 11 can be evaluated using $\Delta s_h/\Delta s_o = \Delta h_h/\Delta h_o$ at $T_{oh}^0$, $\sim 0.25$ [9] so that $(s_h - s_o) \sim 0.75\Delta s_o$. Assuming that $\Delta s_o$ is the same as at atmospheric pressure, which for a heat of fusion of 280 J/g at 415 K is $280/415 \cong 0.67$ J/g K, gives $(s_h - s_o) \cong 500 \, \text{J/kg K}$, whence

$$(T^0_{oh} - T_{oh}) \cong (1/250\lambda)(\sigma_{eo}/\varrho_o - \sigma_{eh}/\varrho_h) \, ,$$

$$(\sigma_{eo}/\varrho_o - \sigma_{eh}/\varrho_h) \cong 250\lambda(T^0_{oh} - T_{oh})$$

and

$$(\sigma_{eo} - 1.08\sigma_{eh}) \cong 2.5 \times 10^5 \lambda (T^0_{oh} - T_{oh}) \tag{12}$$

in SI units.

Taking the figures cited above [24], a depression $\geq 3 \, \text{K}$ for lamellae $\sim 15 \, \text{nm}$ thick when substituted in Eq. 12 with $\sigma_{eo} = 93 \, \text{mJ/m}^2$, places $\sigma_{eh} < 76 \, \text{mJ/m}^2$. This is still a comparatively high figure but it would be reduced if further experiments, at lower annealing temperatures, showed that the hexagonal phase had formed.

## 4.3
## Morphology

Polymers are unique in the extent of the detail of their history which they retain in their morphology, essentially because of the restricted mobility of long molecules once added to a lamella. Indeed polymer morphology has driven almost all advances in understanding the fundamental nature of polymeric self-organization, not least chainfolding. In the present case it demonstrates clearly that polyethylene lamellae crystallized at atmospheric pressure did not have a hexagonal precursor.

Such lamellae are single crystals with habits reflecting the orthorhombic symmetry of the lattice. As Fig. 2a shows there is no sign of a central nucleus with a habit of higher symmetry appropriate to the hexagonal phase. When lamellae are formed in the hexagonal phase then transform to the orthorhombic they invariably form complex threefold twins [24]. This is clearly evident in the circular lamellae of Fig. 2b which are now subdivided into regions of three orientations identified by lines mutually inclined at $\sim 60°$ which are the traces of the shear planes used to incline molecules to lamellar normals in the lower symmetry structure. Similarly, electron diffraction shows [24] that when lamellae are annealed into the hexagonal phase then returned to the orthorhombic, their initial single crystalline nature is lost and invariably transforms to a threefold twin. The reason is that any of the three equivalent $\langle 10.0 \rangle$ lattice vectors in the plane of a hexagonal lamella can transform to the $b$ axis of the orthorhombic cell, the other two becoming $\langle 110 \rangle$ vectors, thereby giving a morphology containing three equivalent possible orientations of the orthorhombic unit cell. It follows very simply, therefore, that the observed single crystals of the orthorhombic phase did not have a hexagonal precursor.

# 5
# Conclusions

There are three principal conclusions:

Insufficiently precise data do not allow the theoretical possibility to be decided whether thin polyethylene lamellae are more stable at atmospheric pressure in the hexagonal phase rather than the orthorhombic though it appears not to be the case.

The single crystalline nature of orthorhombic polyethylene lamellae shows simply and clearly that they did not have a hexagonal precursor. Had they done so they would have been threefold twins.

The orthorhombic and hexagonal phases of polyethylene crystallize independently in accordance with the phase diagram and kinetic competition during growth.

## References

1. Mitchell GR, Rosi-Schwartz B, Ward DJ (1994) Phil Trans R Soc A 348:97
2. Imai M, Mori K, Mizukami K, Kaji K, Kanaya T (1992) Polymer 33:4451
3. Strobl G (2000) Eur Phys J E3:165
4. Keller A, Hikosaka M, Rastogi S, Toda A, Barham PJ, Goldbeck-Wood G (1994) Phil Trans R Soc A 348:3
5. Shabana HM, Guo W, Olley RH Bassett DC (1993) Polymer 34:1313
6. Geil PH, Anderson FR, Wunderlich B, Arakawa T (1964) J Polym Sci A 2:3707
7. Hatekayama T, Kanetsuna H, Hashimoto T (1973) J Macromol Sci B 7:411
8. Bassett DC, Turner B (1972) Nature Phys Sci 240:146
9. Bassett DC, Turner B (1974) Phil Mag 29:925
10. Bassett DC, Block S, Piermarini GJ (1974) J Appl Phys 45:4146
11. Bassett DC (1982) The crystallization of polyethylene at high pressures. In: Bassett DC (ed) Developments in crystalline polymers I. Appl Sci Lond, p 115
12. DiCorleto JA, Bassett DC (1990) Polymer 31:1971
13. Bassett DC, Khalifa BA, Turner B (1972) Nature Phys Sci 239:106
14. Bassett DC, Turner B (1974) Phil Mag 29:285
15. Pechhold W, Liska E, Grossman HP, Hagele PC (1976) Pure Appl Chem 96:127
16. Patel D, Bassett DC (2002) Polymer 43:3795
17. Jackson JF, Hsu TS, Brasch JW (1972) J Polym Sci B 10:207
18. Bassett DC (1981) Principles of polymer morphology. Cambridge University Press, Cambridge, p 179
19. Rees DV, Bassett DC (1971) J Polym Sci A-2 9:385
20. Keller A, Hikosaka M, Rastogi S, Toda A, Barham PJ, Goldbeck-Wood G (1994) J Mat Sci 29:2579
21. Abo el Maaty MI, Bassett DC (2001) Polymer 42:4957
22. Pennings AJ, Zwijnenburg A (1979) J Polym Sci Polym Phys Edn 17:1011
23. Vaughan AS, Ungar G, Bassett DC, Keller A (1985) Polymer 26:726
24. Khalifa BA, Bassett DC (1976) Polymer 17:291

Adv Polym Sci (2005) 180: 17–44
DOI 10.1007/b107231
© Springer-Verlag Berlin Heidelberg 2005
Published online: 29 June 2005

# Analysis and Observation of Polymer Crystal Structures at the Individual Stem Level

Bernard Lotz

Institut Charles Sadron, CNRS - ULP, 6, rue Boussingault, 67083 Strasbourg, France
*lotz@ics.u-strasbg.fr*

**Abstract** Several theories or schemes of polymer crystallization that differ from classical nucleation and growth processes have been put forth recently. They assume some form of structure development in the polymer melt prior to crystallization. This ordering assists crystallization or initiates the build-up of crystal precursors that ultimately form the full-grown crystal by accretion and reorganization. These schemes are evaluated by analysis of the resulting crystal structure (by adopting a strictly structural standpoint). More precisely, the outcomes of selection processes that take place during crystallization of syndiotactic and isotactic chiral but racemic polyolefins are visualized. Both types of polymers can form right-handed or left-handed helical stems in a crystal lattice (in other words there is a conformational choice), but the hand of each helical stem must obey the crystallographic symmetry rules corresponding to the phase (either chiral or antichiral) that is produced. Direct observation of the helical hand of stems building up a single layer and embedded in their crystallographic environment is not normally achievable. It can however be approached using a combination of epitaxial crystallization on a foreign substrate and Atomic Force Microscopy (AFM). Indeed, selective dissolution of the substrate makes it possible to reach the first layer deposited on that substrate and image it (for example by Atomic Force Microscopy). The stems that build up isochiral layers can be shown to have a common helical hand. In one favorable case (the fully antichiral crystal form I of syndiotactic polypropylene), the hand of *individual* stems has been determined. These observations and analyses indicate that the helical hand of stems is highly dependent on the substrate or growth face topography; in other words they indicate that the depositing stem probes and adapts to the surface structure prior to successful attachment. These observations strongly support a crystallization process controlled by the growth front rather than by earlier events that may take place in the polymer melt. In a different approach, use of polyolefins that bear a chiral side-chain and adopt preferred helical conformations in solution and in the melt has been suggested as a way to investigate the relationship (if any) between the helical hand in the melt and that of the resulting crystal structures. On cooling from the melt, they form liquid crystalline phases that later convert to the final crystal structure. In one documented case at least, the final crystal structure is antichiral, whereas the liquid crystalline structure is chiral. These systems, although highly specific and possibly not representative of more common polymers, provide an opportunity to investigate molecular processes that may take place if some type of preordering takes place in the polymer melt.

**Keywords** Epitaxy · Nucleation and growth processes · Helical polymers · Chiral polyolefins

# 1
# Introduction

The molecular processes that take place during polymer crystallization can be investigated from different viewpoints. A global approach determines the shape and conformation of chains in the polymer melt, and infers processes that lead to the lamellar chain folded crystal. Alternatively, analysis of the final crystal structure may help us to work backwards and gain insights into processes that take place in the last stages of crystal build-up. It is clear, however, that the two approaches should lead to the same conclusions, since the crystallization processes that occur should not, and certainly do not, depend on the viewpoint adopted for their analysis.

This has not been so in the past. Indeed, these two approaches have led to diverging opinions, notably regarding the structure of the amorphous layer (chain folds) and its organization – which obviously has a strong bearing on the crystallization process itself. Recall that Flory, starting from his analysis of the chain conformation in the polymer melt, could not imagine how the random coil chain would adopt a (more or less) regularly folded chain conformation. In particular, he considered that the folds on the lamellar surface must be of the "switchboard" model; in other words the folds are *not* globally oriented relative to the underlying crystal lattice or growth faces. Conversely, his more structure-oriented colleagues (Keller, Frank, and so on) had access to local investigation techniques (electron microscopy, electron diffraction) and so were more sensitive to indications of local order, even in the fold surface. For example, the sectorization of single crystals observed in dark field electron microscopy strongly suggests different fold orientations in the fold planes of the different sectors. This debate is well summarized in the proceedings of a landmark meeting of the Faraday Society [1]. Meanwhile, the case of *non*-random orientation of the folds, even for bulk crystallization, has received strong support from the so-called "polymer decoration" technique [2].

The technique rests on the crystallization of polyethylene vapors produced under vacuum and condensed on the fold surface of polyethylene and other polymers. The condensed chains become aligned parallel to the folds, and interact with them through short-range interactions (van der Walls forces). They produce small stacks (rods) of chains normal to the fold orientation. The rods do have preferred orientations, which indicate that the outermost surface of the crystal – the folds – are oriented roughly parallel to the macroscopic growth front and are by no means oriented randomly. The polymer decoration technique, since it rests on short-range interactions, is less or not efficient when dealing with polymers that bear longer side chains, since the conformational freedom of the latter "hides" the fold orientation. However, Atomic Force Microscopy (AFM) in the lateral force mode is able, by adjusting the tip force, to "feel" a preferred fold orientation, even when the folds are burried beneath the top surface. Using this approach, a sectorization (a preferred fold orientation) has been evidenced in (for example) single crystals of poly(4-methyl-pent-1-ene) (P4MP1) even though the top surface is encumbered with side-chains [3]. If anything, this earlier debate and its outcome should teach us that information provided by very local scale indicators – the fold, the crystalline stem – should not be overlooked by any form of analysis of the crystallization process as a whole.

Oddly enough, a very similar debate has been going on in recent years. It has been triggered by a string of new results, many of which are summarized in parallel contributions in this issue. As a result, novel theories or schemes for polymer crystallization have been put forward. They assume some form of preordering of the polymer melt prior to crystallization. In this view, polymer crystallization rests on molecular processes that are more or less directly derived from the melt structure. Two major processes have been considered: (a) spinodal decomposition or spinodal assisted crystallization leading to densification or gelification prior to crystallization, as advocated by, say, Kaji, Olmsted, McLeish and others [4–6], or (b) clustering of chains (or more precisely, of the future crystalline stems) in a kind of loosely organized bundle, condensation of bundles and "crystal perfection" of granular crystalline layers that ultimately results in the lamellar crystal as we know it, as advocated by Strobl [7].

Doubts about the impact on crystallization of such processes have already been raised by the present author in a paper that, very gracefully, Gert Strobl allowed to be published in parallel with his own contribution that presented a different viewpoint [8]. In that paper, the preeminence of a more classical "nucleation and growth" scheme (Fig. 1) was advocated: crystallization is viewed as a more *sequential* process in which incoming stems probe the crystal growth face and are accepted if they fulfill the correct criteria. If not, the incoming stems are rejected or must undertake conformational adjustments. In other words, the classical nucleation and growth process can be seen as dominated or controlled *by the crystal* (substrate structure, or

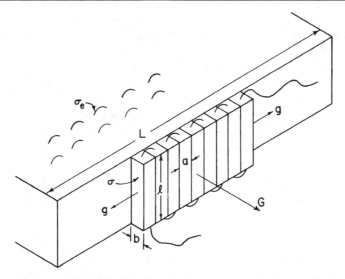

**Fig. 1** The conventional (and idealized) nucleation and growth scheme of polymer crystallization as illustrated by Hoffman and Lauritzen. Growth is a sequential process with successive deposition of stems on a crystalline substrate, here the polymer. The lateral dimensions and length of the stem are represented by $a$, $b$ and $l$; $G$ and $g$ are the overall growth rate and lateral spread on the growth face, $\sigma$ and $\sigma_e$ are the lateral and end surface energies. Substitution of the polymer substrate by a low molecular weight organic material allows observation (via AFM) of the first layer deposited on the substrate (of the depositing strip represented here). Reproduced from [15] with permission

inside-out selection process) whereas the more recent theories suggest that *the melt* (pre)organization or structure governs the crystallization process (outside-in).

In order to settle these divergences, it is necessary to determine whether the initial (melt) or the final (crystal) stage takes precedence in the crystallization process. A purely structural approach can be envisaged, and is exclusively developed here. *If* it is possible to demonstrate that the structure of the crystalline substrate imposes its own crystallographic rules (that are different from the melt), nucleation and growth prevails. *If* (and if so to what extent?) the "pre-ordering" of the polymer melt is transferred to the crystal (or at least leaves a trace in the final crystal structure), the melt organization process becomes a major ingredient.

In this structural approach, and as already detailed in the paper referred to earlier [8], two major criteria may be used to characterize the building blocks of the lamellar crystals, namely the stems. They are the stem length and, for helical chiral but racemic polymers, the helical hand.

The stem length issue has already been discussed in an earlier contribution [8]. The stem length *is* important, but it is a relatively "blunt" criterion. On the one hand, the initial length is not discriminatory enough to impair attachment. Mistakes are tolerated, which slow down but do not stop the crystallization process, as demonstrated by the so-called "surface poisoning" observed for paraffins and low M.W. poly(oxyethylenes), and its subsequent "healing" when "incorrect" stem lengths change to smaller or larger values that are exact sub-multiples of the chain length [9]. On the other hand, the length (or its ultimate value) is the same (allowing for some fluctuations) for all the stems in the crystal, or at least for all stems in any given growth sector.

The analysis of helical hand turns out to be a major tool when investigating the crystal structure of chiral but racemic polymers (such as isotactic polypropylene) that can adopt either right-handed or left-handed helix conformations. Indeed, right or left helical hand is a conformational "tag" attached to each and every stem included in the lamellar crystal, and is a permanent memory of the crystallization of that stem. As such, it provides very local information on the crystallization process, specifically at the level of individual stems. When the right-handed or left-handed helical sense option exists (for isotactic and syndiotactic polyolefins), the helical hand must be selected to conform to the crystal modification generated: chiral (made of helices of only one hand) or antichiral or racemic (made of right-handed and of left-handed helices). Furthermore, in the racemic case, the sequence of helical hands is imposed by the crystallographic symmetry, which suggests that the structure of the growth front has precedence in the deposition process.

Plenty of information on the helical hand can be accessed by analyzing the crystal phase structure (unit-cell symmetry). For example, as developed earlier [8], a very intimate knowledge of the helical hand in the different structures of isotactic polypropylene (iPP) is available ("racemic" $\alpha$ and $\gamma$ phases, or chiral $\beta$ phase). In the specific case of $\alpha$ iPP, both the helical hand and azimuthal setting of each and every stem can be read from the so-called "quadrite" morphology, a unique branched lamellar morphology generated by a homoepitaxy specific to this phase [10]. Further analyses dealt with the $\beta$ and $\gamma$ crystal structures of iPP and with the structures of stereocomplexes of, say, poly(L-lactide) and poly(D-lactide), for which a specific sequence of stems that are part of different molecules must be fulfilled. It was argued that the underlying selection of helical hands (or, for the stereocomplexes, selection of stems of different molecules) is hardly compatible with a simple condensation of a pre-structured polymer melt [8].

As a continuation of the earlier and very lively forum experience (that included the two mentioned papers and two commentaries [11, 12]), the present contribution develops some further arguments based on helical hand of crystalline stems in favor of a nucleation and growth crystallization scheme. Specifically, the crystallization process is analyzed by taking two additional approaches:

- In Sect. 2, we consider an approach even more direct than in the preceding contribution, and a length scale even more local. We take advantage of our mastering of *epitaxial crystallization* on low molecular weight crystalline substrates and our analysis of the epitaxially-crystallized film structure by electron diffraction and Atomic Force Microscopy (AFM) to analyze – again – helical hands of crystalline stems. In epitaxial crystallization, the substrate crystalline structure is of course different from that of the polymer itself. As developed later, epitaxial crystallization is nevertheless a very good approximation to the real growth process.

- In Sect. 3, we suggest that investigation of polymer crystallization processes may be aided by using polymers that have helical conformations, and a preferred hand, in the melt or solution. Such polymers have been observed since the early days of the stereospecific polymerization of polyolefins [13]. They bear a side chain with a chiral carbon atom and experience a macromolecular amplification of chirality [14]. Their main chain therefore adopts a helical conformation with a preferred helical hand, as indicated by a significant increase (and sometimes reversal) of the polymer optical rotary power compared to that of the monomer. This preferred helical hand may be compared with the helical hand(s) existing in the final crystal structure(s). While the present investigation, performed with colleagues in Naples and Zurich, is still under way, the relevance to the present debate is obvious since the precursor liquid crystal phase is likely to possess many of the features supposed to characterize the pre-ordering in the polymer melt.

Before developing these various topics, it may be worth indicating the spirit in which this contribution is presented. It is not meant to advocate old-fashioned theories of polymer crystallization beyond reason. It merely attempts to remind the reader of a number of experimental facts that recent theories may have a hard time explaining, and that should not be overlooked. It should therefore be considered merely as a "note of warning", to use the words of the late H.D. Keith, about the earlier contribution [8]. However, since at times it develops controversial issues or interpretations, the examples used to support these interpretations are taken exclusively from works with which the present author has been or is associated (but the co-authors are in no way responsible for the interpretations developed in the present context!). Last but not least, this approch avoids the risk of mis- or over-interpreting data of colleagues whose works were not meant to be dragged into this type of debate.

# 2
# Analysis of Helical Hand Selection in Epitaxially Crystallized Films

## 2.1
## AFM and Electron Diffraction on Epitaxially Crystallized Layers

Let us consider Hoffmann and Lauritzen's classical "nucleation and growth" scheme of polymer crystallization that can be found in many textbooks (Fig. 1) [16]. Without necessarily adhering to the simplification assumed in this scheme (which would correspond to an ideal, so-called Regime I crystallization process), one must note that the depositing chain will ultimately be squeezed between the substrate and the molecular strips that will be subsequently deposited onto it – it will ultimately be embedded in a crowd of the same material. In other words, there is no way to see the stems that have deposited in Fig. 1, other than by relatively indirect and in any case global methods. These methods mostly use some way of tagging one chain to differentiate it from its neighbors: use of deuterated chains has enabled the chain trajectory to be investigated by neutron scattering [16] or infrared spectroscopy [17].

Epitaxial crystallization combined with Atomic Force Microscopy (AFM) and electron diffraction provide a means to overcome this difficulty. Indeed, in epitaxial crystallization, the polymer substrate in Fig. 1 is replaced by a low molecular weight substance that can be dissolved away with a suitable solvent. In doing so, the deposited strip becomes exposed, and can be seen using local probe techniques – AFM. This observation is aided by the fact that low molecular weight substrates usually form large, flat crystals, which, once dissolved, are highly suited for AFM examination. AFM makes it possible to visualize the very first growth layer deposited on a growth front – in favorable cases down to the helical hand of individual stems.

Obtaining an image of the first layer deposited on a foreign substrate can be rightly considered to be not completely representative of bulk crystallization. However, this is only the first layer of a thin polymer film, the structure of which can be investigated by electron diffraction. The two techniques are indeed very complementary. AFM probes the first layer, whereas electron diffraction determines the structure of the thin film as a whole – the structure of the film interior.

Electron diffraction therefore makes it possible to establish that the structural continuity of the film is ensured. For example, it can differentiate the chiral and the racemic crystal polymorph of a given polymer (it can tell if the selection of helical hands observed in the first layer is still operative in layers deposited subsequently, away from the foreign substrate). As such, electron diffraction probes growth processes taking place in the polymer itself, as opposed to growth on a foreign substrate. Recall that deposition of, say, polyolefin stems is governed mainly by van der Waals forces that fade away beyond some 7 to 10 Å. For isotactic polypropylene, the third layer is 10 Å away from the substrate, and barely knows of the existence of the foreign substrate. Deposition of the fourth layer can already be considered to be representative of bulk crystallization. Bearing in mind that thin films used for electron diffraction range from a few tens to ≈ 100 polymer layers, electron diffraction clearly helps determines features of bulk crystallization. In the present context, it establishes whether the selection processes operative in the first, epitaxially-deposited layer remain operative in the next layers – indeed, if the AFM images of the first layer are also representative of the crystallization of the bulk as a whole.

## 2.2
### Structural Selection Rules Governing Epitaxy of Polymers

Epitaxial crystallization of polymers has been investigated for a wide variety of substrates: minerals (alkali halides, talc, mica, and so on), low molecular weight organic materials (condensed and linear aromatics, benzoic acid and many of its substituted variants and their salts or hemiacids, other organic molecules of different types), and other crystalline polymers.

Epitaxial crystallization utilizes some form of lattice (dimensional) and structural (surface topographies) matching of the deposit (usually the polymer) and the substrate crystal. This matching has been demonstrated in a number of ways in the field of polymer crystallization [18] and only the general rules are (briefly) recalled here.

1. In many cases, and especially for polymers with a "stretched out" conformation (such as the planar zigzag of polyethylene), dimensional matching of the interchain distance with some substrate periodicity is the major criterion for polymer epitaxy. Depending on the substrate periodicity, different crystal planes of the stable crystal phase can become contact planes. In polyethylene, the (110), (010) and (100) contact faces match substrate periodicities of ≈ 4.5 Å, ≈ 5 Å and ≈ 7.5 Å, respectively. When the substrate periodicity is too different from the interchain distances that exist in the stable crystal phase, unstable crystal modifications may be produced. For polyethylene, the monoclinic phase was obtained with three different contact planes with interchain distances of ≈ 4 Å, ≈ 5.5 Å and ≈ 9.5 Å, respectively [19]. A very similar situation exists for isotactic polypropylene that can be obtained either in its antichiral α phase or in its chiral β phase (both phases based on the standard three-fold helix conformation of iPP) by using specific nucleating agents [20–22].
2. For polymers that can exist in different crystal structures based on different chain conformations, epitaxial crystallization can induce these various crystal structures (it can impose different chain conformations). The most

illustrative example is provided by isotactic poly(1-butene) (iPBu). iPBu could be obtained by epitaxial crystallization in Form I (racemic, three-fold helix) [23], Form II (racemic, $11_3$ helix) and Form III (chiral, four-fold helix) [24]. The case of Form I will be further developed later on. However, it is clear from this enumeration that the incoming chain, especially if helical, adapts to the substrate structure, and that the depositing layer conforms to the constraints set by the substrate (periodicity and, as seen later, chirality and orientation).

3. Epitaxial crystallization of helical polymers may involve three different features of the polymer chain or lattice. These are: (a) the interchain distance (as for stretched out polymers), (b) the chain axis repeat distance, and (c) the interstrand distance – the distance between the exterior paths of two successive turns of the helix. The two former periodicities are normal and parallel to the chain axis direction, and are therefore not usually sensitive to the chirality of the helix (unless the substrate topography is asymmetric and favors a given helical hand). However, the interstrand distance is oblique to the helix axis (it is normal to the orientation of the outer chain path) and therefore has different, symmetric orientations relative to the helix axis for left-handed and right-handed helices (Fig. 2). In other words, epitaxies that involve the interstrand distances are discriminative with respect to helix chirality. This discrimination becomes visible if the crystal structure is based on whole layers of isochiral helices. Such a situation does indeed exist for isotactic poly(1-butene), Form I, that will be considered soon.

Some of our earlier results on AFM and electron diffraction of epitaxially-crystallized thin films are briefly recalled in the next subsections, and are

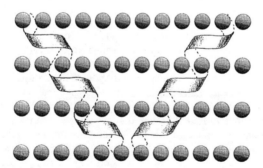

**Fig. 2** Epitaxy of right-handed and left-handed helices on a substrate that matches the interstrand distance. The rows of dots represent surface features of the substrate (e.g. rows of chlorine atoms of p-chlorobenzoic acid), and deposition of helices is seen from the substrate side through this surface layer. The deposition of the first layer of helices is selective with respect to the helical hand, since antichiral helices must lean to opposite directions in order to match the substrate periodicity. Reproduced from [21] with permission

discussed in the light of the "selection of helical hand" issue, in other words in the broader context of mechanisms of polymer crystallization. Three cases are considered: isotactic poly(1-butene) in Form I and syndiotactic polypropylene in Forms I and II (both with $T_2G_2$ chain conformation). AFM results are emphasized, electron diffraction results being discussed only when relevant. As will become apparent, the AFM and electron diffraction analyses are consistent, which indicates that the selection of helical hand observed for the stems in the first layer is indeed representative of the bulk crystallization.

## 2.3
## Epitaxial Crystallization of Isotactic Poly(1-Butene), Form I

The crystal structure of isotactic poly(1-butene), Form I, is based on a trigonal unit-cell with R3c symmetry. The unit cell houses six chains, three right-handed $3_1$ helices, and three left-handed $3_2$ helices. Right-handed and left-handed helices are associated in bilayers parallel to the (110) plane of the unit cell (Fig. 3a) [25]. Furthermore, the right-handed and left-handed helices are oriented in opposite azimuthal directions (in Fig. 3a, one of their side-chains is oriented either towards the top or towards the bottom of the page). Due to this arrangement, all flat faces of the three-fold helices (seen in chain axis projection) are located in the same plane, (110). Furthermore, the flat faces that are in contact in the (110) plane are different: one exposes only right-handed helices, and the other exposes only left-handed helices. When these faces are viewed from the side, only the side chains, and more specifically one $CH_2 - CH_3$ side chain and the end $CH_3$ group of the next side chain are visible. Their tilt to the helix axis is similar to that of the main chain: the structure of the exposed face *indicates the helical hand of the underlying main chain* even though the latter is not accessible to AFM imaging.

**Fig. 3** (**a**) Crystal structure of isotactic poly(1-butene) (Form I′) as seen along the chain ▶ axis. Right-handed and left-handed helices are shown in cylinders, and balls and sticks, respectively. Note that, when exposed, any one (110) plane (horizontal, parallel to the long bisector of the unit cell) is populated with side chain groups $CH_3$, $CH_2$ and $CH_3$ attached exclusively to right-handed helices, or to left-handed helices (**b**) Atomic Force Microscope image of the (110) contact face of epitaxially crystallized iPBu Form I′, and illustration of the methyl and ethyl groups that are imaged. Only one of the two sets of lamellae 24° apart has been selected. About 14 chains are imaged in this view (chain axis is horizontal). The tilt of the side chains indicates that all of the helices in this layer are left-handed. Reproduced from [21] with permission (**c**) Electron diffraction pattern of an epitaxially crystallized thin film of isotactic poly(1-butene) (Form I′). The zone selected comprises only one chain axis orientation (here, the chain axis is vertical). Note that the pattern is asymmetric, which indicates that in spite of its multilamellar structure, the film is truly single crystalline. The strong reflection on the second layer line corresponds to the prominent planes imaged in AFM and indicates that the first layer deposited is made of left-handed helices, as in the AFM image. Reproduced from [24] with permission

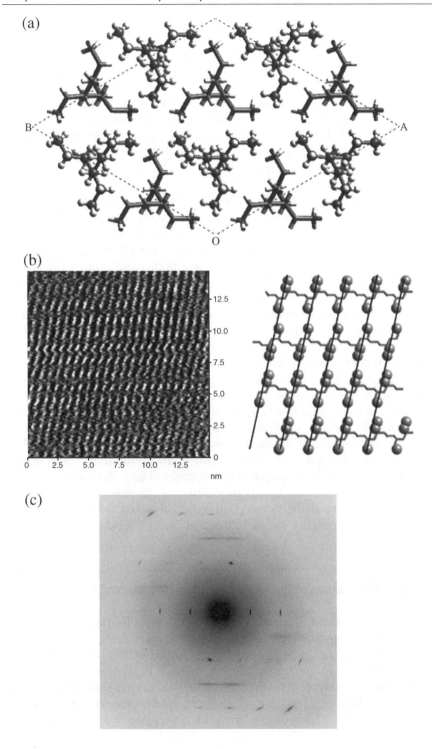

Epitaxial crystallization of iPBu, Form I, on a substrate that matches the interstrand distance produces a film in which two sets of lamellae are at an angle of 24°, as is expected from the 12° tilt of the side groups to the helix axis normal [24]. The discrimination of helical hands illustrated for a single chain in Fig. 2 does indeed take place in the whole epitaxially crystallized first layer. When zooming in on only one of these sets of lamellae, and reaching a resolution of less than ≈ 5 Å, AFM images (Fig. 3b) show the rows of methyl and ethyl groups that build up the contact face [26]. However, the resolution is not sufficient to differentiate the exposed methyl versus ethyl groups, which would have made it possible to determine the syncline or anticline orientation of the helices in the layer.

The AFM images actually show that the selection of helical hands has been flawless: all helices (even if not resolved individually) are left-handed in this first layer. In the portion of the film imaged in Fig. 3b and in this specific set of lamellae, no right-handed helices are imaged, since they build up the second layer away from the substrate (the layer immediately under the imaged one), and all even layers subsequently (with the left-handed helices in all odd-numbered layers), in agreement with the crystal structure of Form I of iPBu1 (all of the helical hands would be opposite had the other set of lamellae been imaged). Evidence for this regular alternation of layers of antichiral helices stems from the analysis of the electron diffraction pattern of the thin film (Fig. 3c). It does indeed indicate that the thin film as a whole has a single crystal texture: the diffraction pattern is not symmetrical, as would be expected for a fiber or even a uniplanar orientation. In particular, the strong diffraction spot on the second layer line indicates the tilt of crystallographic planes containing the side-chains illustrated in Fig. 3b. This tilt is the same for the right and left-handed helices in this layer (helices are related by a glide plane parallel to the layer), again in agreement with the crystal structure of iPBu-1. It also implies that the first, epitaxially crystallized layer deposited on a "foreign" substrate has a structure representative of the bulk of the material. In particular, the very stringent selection of helical hands observed in the first layer is equally operative for all subsequent layers (in other words for bulk crystallization).

## 2.4
### Epitaxial Crystallization of Syndiotactic Polypropylene, Isochiral Form II

The second and third illustrations of epitaxial crystallization deal with syndiotactic polypropylene (sPP). Syndiotactic polymers are by design susceptible to forming either right-handed or left-handed helices, and are therefore suitable materials in the present context of helical hand selection.

Syndiotactic polypropylene exists in various crystal modifications. The most stable chain conformation is helical and involves a succession of the type TTGG or TTG⁻G⁻. The helix conformation is very nearly a rectangular staircase on two sides of which one side-chain $CH_3$, one main-chain $CH_2$ and the next side-chain $CH_3$ form three steps of the staircase. Indeed, they form a row aligned at $\approx 45°$ to the helix axis, and are contained in a plane parallel to (100) (*bc* plane) (Fig. 4a). This chain conformation is found in two different crystal structures, Form I and Form II. Both forms are made of structurally well characterized sheets of helices parallel to the *bc* plane. Form I is fully antichiral, each helix on the rectangular lattice being surrounded by four helices of opposite hand. Form II is chiral, and differs from Form I by a halving of the *b* parameter of the unit-cell, and a *b*/2 shift in successive layers [27]. (Fig. 4b and c). In the present context, we simply need to note that two different layers, one made of antichiral helices and one made of chiral helices, can exist. Their packing energies cannot be significantly different. As an illustration, the chiral Form II becomes the stable form when crystallization takes place under high hydrostatic pressure [28], above about 1.5 kbar, as established in a pressure-temperature phase diagram [29]. Also, packing defects are introduced in the antichiral Form I upon crystallization at low temperatures that correspond to disruption of the full antichiral packing by incorporation of neighboring isochiral helices [30, 31], (Fig. 4d).

Both Form I (considered later) and Form II can be obtained by epitaxial crystallization. At atmospheric pressure, the unstable, chiral Form II can be "forced" to crystallize by using an appropriate substrate, namely 2-quinoxalinol, as assessed by the electron diffraction pattern (Fig. 5a) (no AFM images are available) [32]. The contact plane is found to be $(110)_{Form\ II}$.

The structural relationship of this face with the substrate (Fig. 5b) illustrates the selection mechanisms that are at play during deposition of each and every incoming stem on a foreign crystalline substrate, and also, by extension, on the growth face of the polymer itself.

The substrate surface is made of parallel rows of bulges that stick out of the contact plane. Only Form II of sPP can match this "hilly" surface. Indeed, the various helices in the (110) contact plane also have bulges, with the bulges of all of the chains in phase along the *c* axis (Fig. 5b). Thus, the topographies of the two contact planes match perfectly along the chain axis of the chiral Form II (although the dimensional fit normal to the chain axis would be better for the (110) plane of the antichiral Form I). For Form I, the bulges of neighboring antichiral chains are out of phase, being shifted by *c*/2. Good matching of the substrate and polymer contact plane topographies is thus impossible. In this case, the steric conflicts that inhibit formation of Form I and the structural matching that favors Form II can be directly deduced from the contact faces' surface structure. The driving force underlying the unnatural isochirality of the polymer contact layer can be clearly identified. It is the surface topography of the substrate layer in epitaxial crystallization – or of the growth front in polymer crystallization- that here dictates the "unnatural" se-

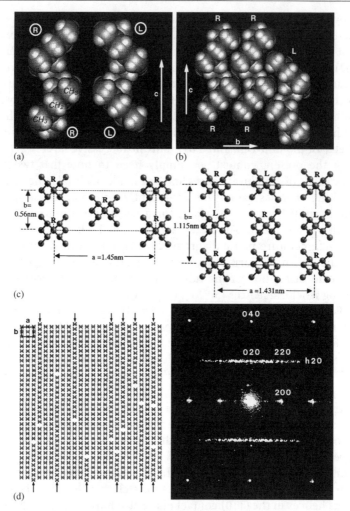

**Fig. 4** (**a**) Conformation of right- and left-handed helices of syndiotactic polypropylene in both Form I and Form II. The helices are seen along the *a* axis, as they are when also imaged by AFM (see Fig. 6). The lateral CH₃, CH₂ and CH₃ groups indicate the orientation of the underlying helix main chain and thus materialize the helical hand. Reproduced from [31] with permission (**b**) The two packing modes of sPP chains in the *bc* layers. A right-handed helix (*center*) is surrounded by a right-handed helix (*left-hand side of the figure*) and by a left-handed helix (*right-hand side of the figure*), which corresponds to packing schemes found in Form II and I, respectively. Note the shift of CH₃, CH₂ and CH₃ groups in the right-hand side antichiral packing, also observed in Fig. 6. Original, unpublished figure produced by A.J. Lovinger (**c**) The unit-cells of sPP: packing of isochiral layers, centered cell (*left*, so-called Form II) and fully antichiral packing (*right*, Form I). Reproduced from [32] with permission (**d**) Structural disorder generated by packing neighboring isochiral helices in an antichiral structure: the *b*/4 shift of layers (*right*) generates a characteristic streak in the optical transform (*left*) and in the *hk*0 diffraction pattern (not shown). Reproduced from [31] with permission

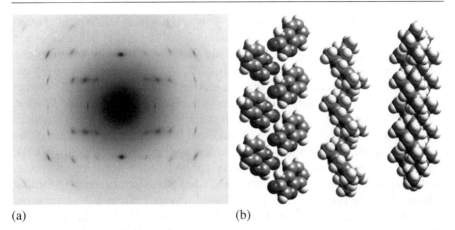

(a)                                               (b)

**Fig. 5** (**a**) Diffraction pattern of sPP crystallized in its chiral Form II on *p*-quinoxalinol. The presence of reflections characteristic of this form and not observed for Form I confirms that the thin film as a whole is mostly in Form II. Chain axis vertical, (110) contact plane. Reproduced from [32] with permission (**b**) Illustration of the topographic interactions that induce Form II rather than Form I: the hilly surface of the 2-quinoxalinol substrate (*left*, seen parallel to the contact face) can only accommodate the Form II (110) plane (*middle*, three chains shown, as seen parallel to the contact plane, chain axis vertical). The (110) plane of Form I (*left*, also three chains represented) has a profile that is not compatible with that of the substrate contact face

lection of a unique helical hand in the depositing layer, an unnatural selection that is further maintained in subsequent layers.

## 2.5
### Epitaxial Crystallization of Syndiotactic Polypropylene, Antichiral Form I

When using a different substrate (*p*-terphenyl), the epitaxial crystallization of sPP yields the *bc* contact plane of the antichiral Form I [33]. This epitaxy provides a unique opportunity to visualize the helical hand of each and every stem in a growth plane of a crystal, a technical prowess achieved so far only for this specific polymer and contact face. The reasons are both conformational and structural. As already indicated, the chain conformation is helical, but the helix has a rectangular shape in *c* axis projection. One side of the rectangle is made of the perfectly aligned two $CH_3$s and $CH_2$, which makes it possible to visualize a fairly big structural entity (a pentane unit seen from the side, about 9 Å long by 4 Å wide, as evaluated from the van der Waals radii). This in turn implies that the AFM tip probes a surface made of helices but a surface that is also perfectly flat. This is at variance with "classical" helices that are circumscribed in a cylinder. In spite of these favorable features, obtaining such an image is not trivial, and is a tribute to Wolfgang Stocker's dedication, patience and expertise in using the AFM.

An unfiltered AFM image of the contact plane is shown in Fig. 6a (area: about 15 nm by 15 nm). It represents a set of parallel helices that are best visualized at grazing incidence along their axis (at about seven to one o'clock). Actually, only one set of helices, all $\approx 11.2$ Å apart, is clearly visible. The helices are left-handed with their stretches of $CH_3$, $CH_2$, $CH_3$ groups oriented at 10 o'clock, nearly normal to the scanning direction, and tilted at about 45° to the chain axis direction. From the $\approx 11.2$ Å periodicity, it is clear that another set of helices exists, intermeshed between these left-handed helices. The set of right-handed helices is less well resolved, since their $CH_3$, $CH_2$, $CH_3$ stretches are oriented nearly parallel to the scanning direction. Nevertheless, the relative shift of the $CH_3$, $CH_2$, $CH_3$ stretches in neighboring helices is clearly resolved in Fourier-filtered images [33], and is as expected from the fully antichiral Form I crystal structure of sPP (Fig. 6b, that represents a small area on the top of Fig. 6a). Also, electron diffraction confirms that the thin film structure is the antichiral Form I.

Figure 6a shows 36 helices, each of which can be assigned a helical hand. It also shows that for each and every one of the 36 helices, the two neighboring helices are antichiral.

What is the probability of forming such a strip of antichiral neighboring helices, knowing that packing of parallel helices is also possible, and is indeed observed? If there were no interactions between the depositing stem and the environment (substrate, neighboring helix), each depositing stem could be either right-handed or left-handed; in other words it would have one chance out of two of adopting the correct helical hand. For 36 chains, there is one chance in $2^{35}$ of achieving the observed alternation of helical hands (one chance in 35 billion). Although this reasoning is, on purpose, strictly limited to the sole stems seen in this AFM image (excluding further lateral growth, and also the underneath layers), it is more than sufficient to demonstrate that a selection process of helical hands was operative.

Did this regular alternation of helical hands result from the deposition process, or was it generated during post-deposition reorganization (in the solid state or near-solid state)? Earlier studies on the structure of single crystals of sPP grown from the melt provide welcome clues. At high $T_c$, the single crystals display sharp reflections that indicate a nearly perfect, fully antichiral Form I crystal structure. The diffraction pattern progressively "deteriorates" as the crystallization temperature decreases: streaks appear that are parallel to $h10$ and $h30$ and become very prominent for low $T_c$s. These streaks indicate a structural disorder whereby some layers (or parts of layers) are shifted by $\approx 2.8$ Å, which is half the span of a helix along the $b$-axis (Fig. 4d). This structural disorder is itself indicative of mistakes in the deposition of helices on the $bc$ face, in other words it is a more macroscopic manifestation of the presence of neighboring isochiral helices in the crystal [31]. Attempts to image such single defects by AFM have been unsuccessful. Nevertheless, their very existence indicates that, when trapped in the crystal, a defect in the alternation

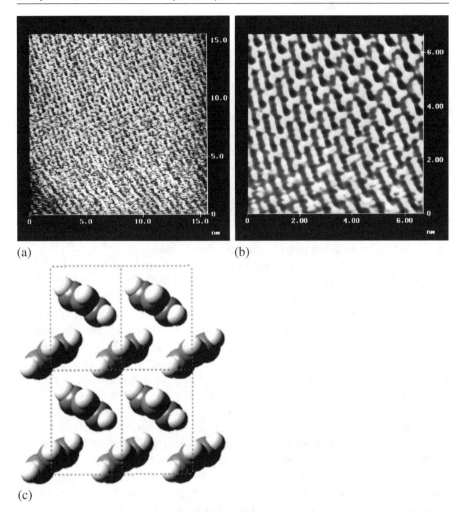

(a)                                        (b)

(c)

**Fig. 6** (**a**) AFM image of the *bc* contact face of sPP in its antichiral Form I, epitaxially crystallized on *p*-terphenyl. Note the strict alternation of right- and left-handed helices (helix axis at one o'clock). Left-handed helices are better resolved than right-handed helices. Reproduced from [33] with permission (**b**) Fourier-filtered image of the upper left area of part (**a**) showing more clearly the right-handed helices that alternate with the left-handed ones. Reproduced from [33] with permission (**c**) The surface structure of *p*-terphenyl used as a substrate for the epitaxial crystallization of syndiotactic polypropylene (parts (**a**) and (**b**)). The top half of the aromatic rings located in the contact plane is shown. Although the dimensional match is acceptable, the chevron-type structure present in this plane does not correspond to that of the contact plane of sPP shown in parts (**a**) and (**b**): no stagger of the molecules, and horizontal periodicity only half of that of sPP in its antichiral Form II

of helical hands is difficult to remove, and leaves a crystallographic "trace" – a local $\approx 2.8$ Å gap in the crystal structure. Conformational and packing energy calculations confirm that helix reversal in sPP is very costly indeed – apparently much more so than for iPP – on account of the deep interdigitation of neighboring helices, almost comparable to that of crankshafts (Fig. 4b and c) [34].

The existence of a structural disorder in sPP generated by fast crystallization conditions, and the fact that it does not "heal" during a later stage after the initial deposition, therefore indicates that the regular alternation of helical hands illustrated in Fig. 6a is the result of a direct deposition process. In any case, generating a regular alternation of helical hands from an initially random sequence of helical hands cannot yield the long range order (at the individual stem scale) demonstrated by Fig. 6a. During reorganization, the system would soon be trapped in numerous local minima of the energy landscape also involving neighboring isochiral chains.

What interactions can explain the regular alternation of helical hands of Fig. 6a? More specifically, is it the $p$-terphenyl substrate structure that forces the strict alternation of helical hands? The structure and surface topography of the contact plane of $p$-terphenyl are known. It was selected because its $ab$ face provides a near-perfect dimensional lattice matching for the $bc$ face of sPP ($a = 8.08$ Å, $b = 5.60$ Å versus $c = 7.4$ Å, $b = 5.6$ Å (or more exactly 11.2 Å$/2$), respectively). The planes of the end phenyl rings that protrude from the contact plane are oblique to the $a$ axis, which generates a kind of shifted herringbone pattern (Fig. 6c), at first sight very reminiscent of the pattern generated by the $CH_3 - CH_2 - CH_3$ stretches in the facing contact plane of sPP (antichiral Form I). This similarity is however misleading. Indeed, the periodicity in sPP is such that neighboring antichiral chains sPP chains interact with every other layer in $p$-terphenyl. As a consequence, the stretches of $CH_3$, $CH_2$, $CH_3$ in successive helices are alternately parallel to the phenyl rings (a probable favorable situation) and nearly normal to the phenyl rings (a much less favorable situation). The structure of the substrate does not dictate the alternation of helical hands observed in Fig. 6a and b. If anything, it would rather call for isochiral helices in the sPP contact plane. The alternation of helical hands therefore results primarily from interactions with the neighboring helices in the growth layer, which implies a nucleation and growth mechanism with probing of the deposition site.

To conclude this section, it may be worth reemphasizing a few conclusions that have been reached.

- The first layer epitaxially crystallized on a substrate of a different nature has many or all of the features of a normal layer produced in bulk crystallization. AFM imaging of the first layer provides the most direct approach that can be imagined to visualize the outcome of the crystallization process. In the present context, AFM is a very local investigation technique, but also a very telling one.

- Electron diffraction used in combination with AFM probes the thickness of the layer (and therefore the true bulk crystallization). Its results are consistent with those of AFM about the organization of helical hands. Both techniques indicate that for iPBu1 and sPP, selection processes are at play for helical hand, that involve the immediate neighborhood of the depositing stem: substrate topography on which the helix is deposited and/or the neighboring helix in the growth front.
- Reversal of helical hands is difficult in a crystalline, and probably also a *pseudo*-crystalline environment (except for helices with long and flexible side chains, as considered in the next section). This difficulty is illustrated by the existence and persistence of structural helix chirality defects in crystals of sPP. Given this difficulty, it would be virtually impossible to reconstruct the perfect alternation of helical hands displayed in Fig. 6a starting from a layer initially made of a more irregular sequence of helical hands.

# 3
# Helix Chirality in the Melt and its Transfer to the Crystalline Solid

The above reasoning regarding helical hand in the crystal rests on the assumption that the polymer melt is either made of random coils, or that, if helical stretches exist in the melt, both right- and left-handed helices exist for chiral but racemic polymers such as isotactic (or syndiotactic) polyolefins. For random coils, the conformation of the incoming chain would simply have to adapt to the crystalline substrate structure. When helical stretches do exist, the sorting-out process described above would have to be fully operative.

A deeper understanding of the crystallization process is desirable and would be achieved if the helical chain conformation in the melt could be determined; more exactly, if it were possible to know which helical hand the chains adopt in the melt, and compare it with that of the crystal that is formed. This issue is now addressed by considering (a) the evidence that the polymer melt is indeed organized to some extent – but this does not require a spinodal decomposition, and (b) the chiral polyolefins mentioned in the "Introduction", for which the helical hand in the melt is known, and for which the helical hand in the crystal is also known.

## 3.1
## Helical Conformations in the Melt?

The existence of helical conformations in polyolefin melts was originally suggested by Garth Wilkes et al in the early 1970s [35]. These authors observed that the X-ray diffraction pattern of the melt of isotactic P4MP1 presents a peak that roughly corresponds to the interhelix distance in the crystal of P4MP1. They concluded that some segments of the chain in the melt must be ordered or helical, or at least that the polymer chain has a cross-section comparable to that of the crystalline state. This may not be totally surprising, given the fact that the densities of the crystalline and amorphous polymers differ by less than 10 to 15% – with the notable exception of P4MP1, for which they are nearly equal. Also, some short-range repetition of *trans*-gauche conformations, imposed by the existence of side-chains in isochiral polyolefins, may explain why (locally at least) the chain axis repeat distance (the average progression of the chain along its path) does not differ significantly from that of a helix. As the interchain distance depends on the square root of the cross-section, it is not surprising that this average interchain distance in the melt bears a reasonable correlation with the interhelix distance in the crystal.

This line of research was followed up by Habenschuss et al, who found similar correlations for a number of polyolefins, ranging from polyethylene to, again, P4MP1 [36]. They note in particular that the peak is more prominent for the chiral polyolefin variant isotactic P3MP1, a point that will be discussed later on. In both the more recent and the older works on this topic, the implicit (or explicit) conclusion is that the polymer melt is organized to some degree.

It is difficult to draw any firm conclusion about the exact state of the polymer in the melt based on this sole observation, but clearer cases can be investigated, as seen next. It is fair to state that in all polyolefins considered so far, even if the melt is structured to some degree (an issue that it is safer to leave open at this stage), the helical hands (if they exist) are equally shared between right and left. If so, the sorting out process developed above should still apply. Conversely, and this argument has already been developed in an earlier paper [8], it is difficult to imagine that a racemic polymer melt structure can yield chiral structures such as the $\beta$ phase of isotactic polypropylene. Admittedly, $\beta$ phase crystals probably have complex structures, made of patches or domains that are antichiral, and each crystal as a whole is a racemic blend of antichiral patches [37]. However, the generation of two-dimensional patches of a sufficient size to yield a diffraction pattern characteristic of the chiral $\beta$ phase requires an ordering of the chiralities of a number of stems that is orders of magnitude larger than the above sPP stems deposited in a single layer of the growth front. Needless to say, in the absence of any selection process, the odds of succeeding in this endeavor would be accordingly lower.

## 3.2
## Helical Conformations in the Melt

In the search for a better system for determining the helical hand of the chains both in the polymer melt and in the crystal structure (in the initial and final stages of crystallization), we have recently become interested in poly-olefins bearing a chiral side chain. Work along this line was performed in the very early days of the development of isotactic polyolefins, and most promi-nently by Pino, in Milan and later in Zurich, but was not followed up in later years [14]. Unfortunately, and probably as a consequence of this loss of inter-est, most of the samples synthesized at that time have been lost. Nevertheless, we obtained a set of samples from Dr. Peter Neuenschwander, who was a col-laborator of Prof. Pino in Zurich [38] Work on these materials is still under way in our laboratory in collaboration with Dr. Buono in Naples. The early re-sults of this study are worth reporting, since they appear highly relevant to the present discussion.

Only two types of polymers are considered here. These are isotactic poly(5-methyl-hexene-1) (P5MH1), with a non-chiral side chain (for the sake of comparison) and mainly isotactic poly(4-methyl-hexene-1) (P4MH1). The side chain of the latter polymer is chiral since the two substituents of the carbon in the $\beta$ position are a methyl and an ethyl group. Polymers that are made only of the S or the R conformers – in other words the true chiral poly-mers (P(S)4MH1 and P(R)4MH1), the racemic copolymer of the (R) and (S) monomers (P(R, S)4MH1) and of course the racemic blend of the two enan-tiomeric polymers – are available.

The crystal polymorphism of the chiral but racemic P5MH1 is, to some extent, very reminiscent of that of isotactic polypropylene. It exists in two crystal modifications. One crystal modification is stable at high temperature, and was observed early on by Corradini et al [39]. Its structure has been redefined as a chiral, frustrated one based on a trigonal cell with three three-fold helices per cell. We have also discovered a second crystal modification produced from solution. It has an orthorhombic unit cell that contains four chains in – again – three-fold helical conformation, for which one must as-sume coexistence of two right- and two left-handed helices. Contrary to the $\alpha$ and $\beta$ phases of iPP, the frustrated structure of poly(5-methyl-hexene-1) is the more stable one [40].

Isotactic poly(4-methyl-hexene-1), with its chiral side chain, is a represen-tative example of a class of polymers that display a so-called "macromolecular amplification of chirality", in the words of Green et al [14]. Mark Green has illustrated this chiral amplification with poly(isocyanates), but acknowledges the early investigations and insights resulting from the pioneering work of Pino and his coworkers on isotactic polyolefins with chiral side chains, in short chiral isotactic polyolefins. For these polymers, the chiral side chain in-duces a slight preference for a given main chain conformation. The slight bias towards one conformation is amplified cooperatively in the main chain, and induces a preferred main chain helical hand. The preferred helicity is man-ifested by unusual optical properties of the polymer melt and solution. The polymer has an optical rotary power (per monomeric unit) that is signifi-cantly larger and sometimes opposite to that of the monomer. The sign of the

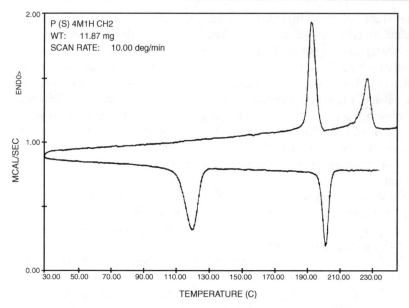

**Fig. 7** DSC melting and crystallization curves of isotactic poly(S)-4-methyl-hexene-1. Note the two melting peaks (at 193.5 and 227.4 °C, respectively, $\Delta H \approx 2.5$ and 1.5 cal/gram) and two crystallization peaks (at $\approx 201$ and $\approx 120$ °C, respectively; same $\Delta H$), as well as the significant temperature gap ($\approx 74$ °C) between the lower crystallization and melting processes. (From [44])

optical rotation helps determine the hand of the helices – for P(S)4MH1, the helices are left-handed.

The crystal structures and indeed the whole phase diagram of P4MH1 are still under investigation, and are not yet fully elucidated. However several interesting results were obtained during the early experiments performed in the 1960s and 1970s. For example, Corradini et al established the crystal structure of P(S)4MH1; it is a tetragonal cell with four chains in $7_2$ helical conformation [41]. Again, the presence of four chains in the cell suggests the coexistence of left-handed and right-handed helices. P(S)4MH1 is therefore a chiral polymer that is *known* to have a preferred or dominant helical conformation in the melt (or solution) and yet crystallizes in an antichiral crystal structure. Moreover, results by Pino and his coworkers [42, 43] on the phase diagram, corroborated by our own results, indicate a complex crystallization and melting behavior (Fig. 7). On heating, two melting peaks are observed at 193 and 227 °C. On cooling P(S)4MH1 first produces a liquid crystalline, presumably chiral phase (peak at 201 °C). Transformation to the stable antichiral phase requires a reversal of helical hand of one stem out of two. Apparently, the stability of the helix form (as a result of the amplification of chirality) delays this transformation process down to about 120 °C. Moreover, the crystal phase is generated by a liquid crystal-crystal transformation, resulting in up to three different orientations of the newly-formed tetragonal unit cell from the parent hexagonal one [44]. These different crystal orientations have already been observed for polyethylene after a transformation from a hexagonal to an orthorhombic cell symmetry [45]. As already indicated in an earlier contribution [8], the existence of different, crystallographically-related unit cell orientations of the final crystal form is probably one of the strongest pieces of evidence indicating a two-step crystallization process, in other words the existence of a pre-ordered phase (presumably liquid crystalline-like).

Preliminary results obtained with the racemic P(R, S)4MH1 (the random copolymer of the R and S enantiomeric monomers) and of the racemic blend of P(S)4MH1 and P(R)4MH1, indicate a more conventional crystallization behavior: crystallization of the tetragonal form takes place in the upper $T_c$ range ($\approx$ 183 and $\approx$ 191 °C, respectively), presumably because both left- and right-handed helices (or helical stretches) are available at that temperature, whereas melting occurs at $\approx$ 211 and $\approx$ 214 °C.

In the debate about existence of pre-ordered states in the polymer melt, as advocated recently, polyolefins with chiral side chains may well become a major investigation tool. Indeed, the macromolecular amplification induces a pre-organization, or at least a preferred helical conformation in the polymer melt or solution. As such, these polymers display very precisely the behavior that is assumed by some of the recent crystallization schemes or scenarios. Furthermore, for the P4MH1 systems considered so far at least, the conformationally racemic character of the stable crystal structure implies that half of the stems must change their helical hands at some stage in the crystallization process – which may greatly delay the formation of this stable crystal structure, as illustrated by P(S)4MH1.

Polyolefins with chiral side chains are systems highly adapted to investigating details of the crystallization process. Indeed, Pino et al have shown that the macromolecular amplification of chirality is actually tunable [46]. For example, substitution closer to the main chain enhances the trend for generating helical structures: P3MH1 exists in a liquid crystalline form up to 400 °C (up to its thermal degradation). Conversely, the tendency to form helices can be softened. Indeed conventional polymers can be induced to form helices of a given sense through copolymerization with a small fraction of a chiral comonomer. The tendency to form these helices can be monitored by adjusting the comonomer composition [47].

Although most of the initial polymers and copolymers are no longer available, these systems should help us to investigate the details and structural consequences of a pre-ordering of the polymer melt, or the formation of a precursor crystal in polymer crystallization. From the present and as yet still sketchy results, it appears however that: (a) if a crystal phase results from the transformation of a liquid crystal phase (usually of hexagonal symmetry), different orientations of the crystal unit cell are likely, which is not observed in conventional polymer crystallization (for example, in polyethylene spherulites the *b* axis is radial, whereas three different orientations would be expected if a liquid crystal phase (even if transient) had been involved), and (b) production of a final crystal phase from a polymer melt that is structured (at least at the level of helical stems) is possible, as confirmed, for instance, by the crystallization of the racemic blend of P4MH1. Moreover, the fact that the racemic polymer and the blend of enantiomers both display only one crystallization peak at a temperature that corresponds to the onset of liquid crystallinity in the chiral polymer suggests that, in the present system, crystallization in the final crystal form may well be triggered or induced by the development of the liquid crystalline phase – which would correspond to the scheme advocated by Strobl. A structural analysis of the resulting phase (one, or several orientations of the unit cell axes, etc) should help establish whether or not a transient phase was at play in the process.


# 4
## Conclusion

This contribution has attempted to show that, although crystallization of polymers is indeed a complex process governed by a number of variables, analysis of the resulting crystal structure suggests that the ultimate step of this process is a relatively simple adjustment of the incoming stem to the crystalline growth front. Indeed, ultimately any crystallization boils down to the sum of very local processes. In polymer crystallization, the elementary building block is the stem that spans the lamellar thickness. For these stems, and for these stems only, the constraints put on successful deposition can be clearly identified. They are mainly geometric or conformational: stem length and, for chiral but racemic polymers, helical hand.

The helical hand of the individual stems in chiral but racemic polymers is a very severe and therefore critical criterion in the crystallization process. The constraints apply for each stem and are dictated by the symmetry of the unit-cell. Contrary to the stem length (which, if incorrect, can be healed or adjusted by later structural reorganization), helix chirality involves a "flip of a coin" type of decision – right- or left-handedness.

On a macroscopic scale, the adjustment of helical hand is an indisputable feature. Indeed, the very fact that different crystal structures (either conformationally chiral or racemic) can be formed from the same initial melt (whether structured or not by, for example, spinodal decomposition) indicates that this selection of helical hand is operative. Even more demonstrative, the occurrence of mistakes in the adjustment of helical hands can have a profound impact on the growth process. For example, the formation of a sufficient patch of isochiral helices in the antichiral $\alpha$ phase of iPP may induce a growth transition to the chiral $\beta$ phase. The rarity of such growth transitions indicates that the selection of helical hands is not only operative, but is also very strict.

On a more microscopic scale, the present contribution has shown that the selection of helical hands can be demonstrated, and actually visualized, by combining appropriate preparation and investigation techniques, namely epitaxial crystallization, Atomic Force Microscopy and electron diffraction. Observation by AFM of individual layers – the first layer deposited on the crystalline substrate – and even, in these layers, of each helical stem, is the closest approach to the selection of helical hand that is conceivable and achievable. This direct observation indicates that the incoming stems can read the helical hand that is set by the substrate (isochirality of the first layer of iPBu1) or can adapt to an unnatural substrate topography by adopting a different crystal modification (isochiral Form II of sPP). The results obtained with sPP crystallized on $p$-terphenyl are even more telling. The strict alternation of helical hands observed in a relatively short stretch of a single layer made of thirty-six stems helps evaluate the very slim chances that any crystal structure displaying some regularity is the result of a rearrangement of an initially disordered structure. This would be the case for precursors (bundles, spinodal decomposition) of the polymer crystal that are formed in the melt, and as such are not submitted to the constraints of crystal symmetry, or have not yet probed the growth face. Conversely, the results presented here reinforce, in our view, the more classical crystallization scheme described as a "nucleation and growth" process. It emphasizes the individuality of the stems as opposed to the macromolecule, as is the case for the more recent schemes.

Finally, we have attempted to evaluate the possible impact of an intermediate liquid crystalline phase and the possibility of "transfer of helical hand information" from the melt to the crystal throughout this process. Assuming that the melt is structured, the melt of chiral but racemic polyolefins would be made of stretches of helical stems that are equally partitioned between left- and right-handed helices. Formation of antichiral structures (such as in $\alpha$ iPP) could be interpreted as indicating a possible transfer of information (but the problem of the sequence of helical hands would still remain). This analysis is, however, ruined by the observation that many of these polymers also form chiral structures (frustrated $\beta$ phase of iPP, Form III of iPBu1). For the achiral poly(5-methyl-pentene-1), the chiral, frustrated phase is actually the more stable one, and can be obtained by melting and recrystallization of a less stable antichiral phase.

A novel approach (or more exactly a revival and extension of an old approach) has been suggested in order to evaluate the transfer of helix chirality information from the melt to the crystal. It rests on the use of polyolefins that bear a chiral side chain. These polyolefins display a chiral amplification which, in the present context, provides us with the possibility of finding out whether helices with a preferred hand exist in the melt, and of pinpointing the precise helical hand. This investigation is still under way and will be developed in the future. Preliminary results obtained in our laboratory corroborate earlier observations by Pino and coworkers [43] who suggested that these polymers form liquid crystalline phases prior to the final crystallization. The structural consequences of such a two-step process can thus be evaluated in the real world. Present results suggest that a two-step process involving a transition or transformation between a liquid crystal phase and a crystal do – again – result in different orientations of the unit-cell of the latter, which is not generally observed in polymer crystallization.

Polymer crystallization is, almost by design, a messy process, given the constraints set by entanglements and so on. However, observation of the resulting crystal structure at a local scale suggests that a significant level of structural order is generated: AFM observations of the first polymer growth strips deposited by epitaxial crystallization on foreign substrates indicate a stringent selection (adjustment) process for the substrate structure. As assessed by electron diffraction, this selection process also takes place in the subsequent layers – that is, in the bulk. The level of organization reached appears to us incompatible with a reorganization of an initially (more) disordered precursor structure. The level of organization appears more compatible with a more classical nucleation and growth process during which the incoming stem probes the substrate surface and structure. Clearly however, the nucleation and growth processes advocated here can and must only be considered as representative ones, or more exactly, the most representative ones that further leave a material trace in the final structure.

To conclude, it must be stated again that the present views derive from a strictly structural analysis of the crystal structure that is generated during the crystallization process, and moreover this analysis has been made at the level of individual stems. As a rule, such a structural analysis is more sensitive to order than to disorder (which is usually manifested at a more global scale, in the form of streaks or broadening of reflections). The local scale analysis developed here may therefore emphasize or overemphasize the structural regularity of the final lamellar structure. In a balanced evaluation of the possible processes contributing to polymer crystallization, this possible bias towards crystallographic order (taken here as supporting evidence for a nucleation and growth process) must be taken into account. It is fully possible that, under appropriate or specific circumstances, different mechanisms are at play – they are certainly at play for deep quench, when the helical hand selection processes can no longer be operative, and which generates for example the smectic phase of isotactic polypropylene. Conversely, and this is the main message of the present contribution, the newer schemes cannot ignore, and must explain the high level of structural (crystalline) order generated by the crystallization process. These schemes have so far mainly emphasized the initial stages of the process, and are indeed compatible with evidence indicating some form of structural ordering in the melt. They have not, however, considered the final stages with as much care, namely the stem-by-stem build-up (or ordering) of the crystal structure. If this ordering is left to as yet little understood reorganization processes in a preexisting structure (involving, for the polymers considered here, reversal and adjustment of helical hands in an initially random, conformationally racemic structure, in addition to the better understood adjustment of stem length), the scenario, or possibly the multiple scenarios of polymer crystallization become much more involved than initially thought, and will require many more investigations. The "nucleation and growth" process appears comparatively simpler and more directly appealing when analyzing, as performed here, the stem organization of the resulting polymer crystal structures.

**Acknowledgements** Special thanks are due to close collaborators, and mainly to C. Straupé, Drs. A. Thierry, J. Ruan and M. Schmutz for their help in various stages of the preparation of this contribution.

# References

1. RSC (eds) (1979) Faraday Discuss 68
2. Wittmann JC, Lotz B (1985) J Polym Sci Pol Phys 23:205
3. Pearce R, Vancso GJ (1998) Polymer 39:6743
4. Olmsted PD, Poon WCK, McLeish TCB, Terrill TCB, Ryan A (1998) Phys Rev Lett 81:373
5. Terrill NJ, Fairclough PA, Towns-Andrews E, Komansheck BU, Young RJ, Ryan AJ (1998) Polymer 39:2381
6. Imai M, Mori K, Mizukami T, Kaji K, Kanaya T (1992) Polymer 33:4457
7. Strobl G (2000) Eur Phys J E 3:165
8. Lotz B (2000) Eur Phys J E 3:185
9. Cheng SZD, Chen JH, Zhang AQ, Barley JS, Habenschuss A, Zschack PR (1992) Polymer 33:1140. See also the contribution by Ungar G, et al. in this issue
10. Lotz B, Wittmann JC (1986) J Polym Sci Pol Phys 24:1541
11. Cheng SZD, Li CY, Zhu L (2000) Eur Phys J E 3:195
12. Muthukumar M (2000) Eur Phys J E 3:199
13. Pino P (1965) Adv Polym Sci 4:393
14. Green MM, Park JW, Sato T, Teramoto A, Lifson S, Selinger RLB, Selinger JV (1999) Angew Chem Int Edit 38:3138
15. Hoffman JD, Davis GT, Lauritzen JI (1976) Hannay NB (ed) Treatise on solid state chemistry, vol 3. Plenum, New York, Ch 7
16. Sadler DM, Spells SJ, Keller A, Guenet JM (1984) Polym Commun 25:290
17. Cheam TC, Krimm S (1981) J Polym Sci Pol Phys 19:423
18. Wittmann JC, Lotz B (1990) Prog Polym Sci 15:909
19. Wittmann JC, Lotz B (1989) Polymer 30:27
20. Yan S, Katzenberg F, Petermann J, Yang D, Shen Y, Straupé C, Wittmann JC, Lotz B (2000) Polymer 41:2613
21. Mathieu C, Thierry A, Wittmann JC, Lotz B, (2000) Polymer 41:7241
22. Mathieu C, Thierry A, Wittmann JC, Lotz B (2002) J Polym Sci Pol Phys 40:2504
23. Kopp S, Wittmann JC, Lotz B (1994) Polymer 35:916

24. Kopp S, Wittmann JC, Lotz B (1994) Polymer 35:908
25. Natta G, Corradini P, Bassi IW (1960) Nuovo Cimento Suppl 15(1):52
26. Mathieu C, Stocker W, Thierry A, Wittmann JC, Lotz B (2001) Polymer 42:7033
27. Lotz B, Ruiz de Ballesteros O, Auriemma F, De Rosa C, Lovinger AJ (1998) Macro-molecules 31:9253
28. Rastogi S, Loos J, Cheng SZD, Lemstra P (1999) Abstracts. 218th ACS Meeting PMSE 218:153
29. Rastogi S, La Camera D, van der Burght F, Terry AE, Cheng SZD (2001) Macro-molecules 34:7730
30. Lotz B, Lovinger AJ, Cais RE (1988) Macromolecules 21:2375
31. Lovinger AJ, Lotz B, Davis DD, Padden FJ Jr (1993) Macromolecules 26:3494
32. Zhang J, Yang D, Thierry A, Wittmann JC, Lotz B (2001) Macromolecules 34:6261
33. Stocker W, Schumacher M, Graff S, Lang J, Wittmann JC, Lovinger AJ, Lotz B (1994) Macromolecules 27:6948
34. Brückner S, Allegra G, Corradini P (2002) Macromolecules 35:3928
35. Wilkes GE, Lehr MH (1973) J Macromol Sci Phys B7:225
36. Kim MH, Londono JD, Habenschuss A (2000) J Polym Sci Pol Phys 38:2480
37. Dorset DL, McCourt MP, Kopp S, Schumacher M, Okihara T, Lotz B (1998) Polymer 39:6331
38. Neuenschwander P, Pino P (1983) Eur Polym J 19:1075
39. Corradini P, Martuscelli E, Montagnoli G, Petraccone V (1970) Eur Polym J 6:1201
40. Buono A, Talarico G, De Rosa C, Thierry A, Lotz B (to be published)
41. Bassi IW, Bonsignori O, Lorenzi GP, Pino P, Corradini P, Temussi PA (1971) J Polym Sci A2 9:193
42. Benedetti E, Bonsignori O, Chiellini E, Pino P (1978) Journées de Calorimétrie et d'Analyse Thermique (JCAPDR) 9A B9:65
43. Bonsignori O, Pino P, Manzani G, Crescenzi V (1975) Makromol Chem Suppl 1:317
44. Buono A, Ruan J, Thierry A, Lotz B, Neuenschwander P (to be published). See also Buono A, Ruan J, Thierry A, Neuenschwander P, Lotz B (2005) Chin J Polym Sci 23:171
45. DiCorleto JA, Bassett DC (1990) Polymer 31:1971
46. Pino P, Ciardelli F, Lorenzi GP, Montagnoli G (1963) Makromol Chem 61:207
47. Carlini C, Ciardelli F, Pino P (1968) Makromol Chem 119:244

Adv Polym Sci (2005) 180: 45–87
DOI 10.1007/b107232
© Springer-Verlag Berlin Heidelberg 2005
Published online: 29 June 2005

# The Effect of Self-Poisoning on Crystal Morphology and Growth Rates

G. Ungar (✉) · E. G. R. Putra · D. S. M. de Silva · M. A. Shcherbina ·
A. J. Waddon

Department of Engineering Materials, University of Sheffield, Sheffield S1 3JD  UK
g.ungar@shef.ac.uk

**Abstract** Recent extensive experimental work and the limited theoretical studies of the phenomenon of self-poisoning of the crystal growth face are reviewed. The effect arises from incorrect but nearly stable stem attachments which obstruct productive growth. Experimental data on the temperature and concentration dependence of growth rates and the morphology of long-chain monodisperse *n*-alkanes from $C_{162}H_{326}$ to $C_{390}H_{782}$ are surveyed and compared to some previously established data on poly(ethylene oxide) fractions, as well as on polyethylene. The anomalous growth rate minima in both temperature and concentration dependence of growth rates are accompanied by profound changes in crystal habits, which have been analysed in terms of growth rates on different crystallographic faces, and in terms of separate rates of step nucleation and propagation. In some cases non-nucleated rough-surface growth is approached. The phenomena covered include "poisoning" minima induced by guest species, the "dilution wave" effect, autocatalytic crystallization, pre-ordering in solution, two-dimensional nucleation, and the kinetic roughening and tilt of basal surfaces.

**Keywords** Polymer crystallization · Nucleation · Long alkanes · Surface roughness · Curved faces

### Abbreviations

| | |
|---|---|
| $A$ | Stem attachment rate |
| $a_0, b_0$ | Unit cell parameters |
| AFM | Atomic force microscopy |
| $B$ | Stem detachment rate |
| $b$ | Width of a molecular chain |
| E | Extended chain form (= $F_1$) |
| $F_2, F_3, ..., F_m$ | "Integer forms" with chains folded in two, three etc. |
| $\Delta F$ | Overall free energy of crystallization |
| $\phi$ | Chain tilt angle with respect to layer normal |
| $\varphi$ | Obtuse angle between (110) and (– 110) planes in alkane and polyethylene crystals. $\varphi/2 = \tan^{-1}(a_0/b_0)$ |
| $\Delta\phi$ | Bulk free energy of crystallization |
| $G$ | Crystal growth rate |
| $\Delta h_f$ | Heat of fusion |
| $i$ | Rate of initiation (secondary nucleation) of a new row of stems on crystal growth face |
| IF | Integer folded |
| $K$ | Slope of the linear dependence of $G$ on $\Delta T$ |
| $L$ | Chain length |
| $l$ | Length of straight-chain segment traversing the crystal (stem length) |
| $l_{SAXS}$ | SAXS long period |
| LH theory | The theory of Lauritzen and Hoffman |
| $m = L/l$ | Number of folds per chain + 1 |
| $M_n$ | Number average molecular mass |
| $n$ | Number of monomer repeat units per chain (e.g., number of carbons in an alkane); also reaction order |
| NIF | Non-integer folded form |
| PE | Polyethylene |
| PEO | Poly(ethylene oxide) |
| $q$ | Modulus of the wavevector, $q = 4\pi(\sin\theta)/\lambda$, where $\theta$ is half the scattering angle and $\lambda$ is radiation wavelength |
| SANS | Small-angle neutron scattering |
| SAXS | Small-angle X-ray scattering |
| $\sigma$ | Side-surface free energy |

| | |
|---|---|
| $\sigma_e$ | End- or fold-surface free energy |
| $T_c$ | Crystallization temperature |
| $T_c^{F_x - F_y}$ | Growth transition temperature between two successive folded forms (e.g., $T_c^{E-F_2}$ is the temperature of transition between extended (E) and once-folded ($F_2$) chain growth) |
| $T_d$ | Dissolution temperature |
| $T_m$ | Melting temperature |
| $T_R$ | Roughening transition temperature |
| $\Delta T$ | $T_m - T_c$ or $T_d - T_c$ = supercooling |
| $v$ | Rate of step propagation on a crystal growth face (often also referred to as $g$) |

# 1
## Introduction

Polydispersity has often been an obstacle to obtaining reliable data on which to base a good understanding of polymer crystallization. The first monodisperse oligomers showing chain folding were long normal alkanes synthesized in 1985 independently by Bidd and Whiting [1] and by Lee and Wegner [2]. Whiting's method has subsequently been perfected and extended to long alkane derivatives by Brooke et al. [3]. Recently, monodisperse oligomers of aliphatic polyamides [4, 5] and poly(hydroxybutyrate) [6] have also been synthesized. Studies on the morphology and crystallization of model polymer systems, including both monodisperse and fractionated linear and cyclic oligomers, have been reviewed recently [7]. A balanced general overview of polymer crystallization theories can be found in [8].

The present article reviews a particular aspect of the crystallization mechanism of long-chain molecules, termed self-poisoning, which has come to light in studies on long alkanes. The effect is closely related to the so-called pinning, which is a key feature of the "roughness-pinning" theory of polymer crystallization. Recent crystallization studies are highlighted. Information from recent and ongoing work adds new detail to the picture of the events at lateral crystal growth surfaces. The issue of kinetic roughness of the basal lamellar surface is also addressed.

## 2
## Quantized Stem Lengths

### 2.1
### Integer and Non-Integer Folding

Monodisperse oligomers long enough to exhibit chain-folded crystallization have been shown to favour integer folding, i.e. such conformations which ensure the location of chain ends at the crystal surface. This applies to all such oligomers studied so far, i.e. long alkanes [9], polyamides [5] and poly(hydroxybutyrate) [6]. It is also prominent in narrow molecular weight fractions of poly(ethylene oxide) (PEO, see Sect. 2.2 and Sect. 4.3.4). Figure 1 lists the observed conformations in a series of solution-crystallized alkanes [9]. Figure 2 illustrates graphically the quantization of lamellar thickness by showing the evolution of the small-angle X-ray scattering (SAXS) curve of $n$-alkane $C_{390}H_{782}$ initially folded in five ($F_5$) during continuous heating. Stepwise crystal thickening through successive integer forms $F_m$ is observed, with the long period $l_{SAXS}$ adopting values equal to $(L/m)\cos\phi$, where $L$ is the extended chain length and $\phi$ is the chain tilt relative to the layer normal. The integer $m$ successively decreases from 5 to 2. Figure 3 shows edge-on lamellae of $C_{390}H_{782}$ on graphite, initially in $F_4$ form, thickened after annealing at 120 °C to $F_2$ and then at 130 °C to the extended-chain form (E) [10].

The $l_{SAXS} = (L/m)\cos\phi$ relationship applies strictly only to solution-grown crystals, while melt crystallization initially produces a transient non-integer form (NIF), where $m$ is a fractional number [11]. However, even in NIF, at least in the case of alkanes, the straight-chain stems do not assume fractional lengths. Where $1 < m < 2$, the crystalline stem length is still exactly $L/2$, but the crystalline layers are separated by amorphous layers containing loose ends (cilia) belonging to chains which are not folded in two but which

| Chain conform. Paraffin | E | F2 | F3 | F4 | F5 |
|---|---|---|---|---|---|
| C102 | + | | | | |
| C150 | + | + | | | |
| C198 | + | + | + | | |
| C246 | + | + | + | + | |
| C294 | + | + | + | + | |
| C390 | + | + | + | + | + |

**Fig. 1** Integer folded (IF) forms observed in long $n$-alkanes $C_{102}H_{206}$ through $C_{390}H_{782}$. For a given alkane more folds per molecule are obtained with increasing supercooling $\Delta T$ (from [7] with permission of American Chemical Society)

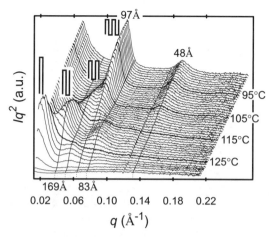

**Fig. 2** Series of SAXS intensity functions recorded during heating of solution-crystallized $n$-$C_{390}H_{782}$, initially in $F_5$ form, and refolding through $F_4$ and $F_3$ to $F_2$ (see symbols) (from [7] with permission of American Chemical Society)

traverse the crystal layer only once. This semicrystalline structure of the NIF has been determined by electron density reconstruction from real-time synchrotron SAXS data [12], backed by real-time Raman LAM (longitudinal acoustic mode) spectroscopy [13] and static [14] and real-time [15] small-angle neutron scattering (SANS) on end-deuterated alkanes. The transient NIF transforms subsequently to either extended-chain (E), once-folded chain ($F_2$) or a mixed-integer folded–extended (FE) form [13] (see Fig. 4).

From the point of view of understanding crystal growth and morphology, the above structure of the NIF means that under any circumstance, whether forming from solution or from melt, the stem length in monodisperse alkane crystals is quantized.

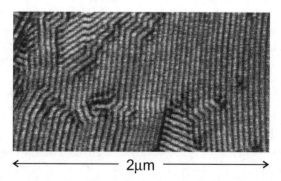

**Fig. 3** Edge-on lamellar crystals of $C_{390}H_{782}$ on graphite, initially in $F_4$ form, thickened after annealing at 130 °C to the extended-chain form (after [10])

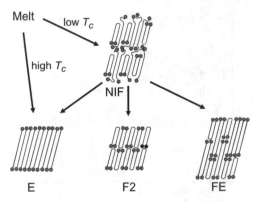

**Fig. 4** Paths to integer folded forms in long alkanes. While solution crystallization gives integer forms (E, $F_2$, ...) directly, for melt crystallization this is true only for the E form. Below the melting point of $F_2$, the transient non-integer form (NIF) appears and subsequently transforms isothermally either to the once-folded $F_2$ form or the mixed-integer folded–extended (FE) form, based on layer triplets. Deuterium- labelled chain ends, prepared for neutron scattering experiments, are indicated with *circles*

## 2.2
### Growth Rate of Crystals with Preferred Thicknesses

In all kinetic theories of polymer crystallization [8] the crystal growth rate $G$ is given by:

$$G = A \left(1 - B/A\right) \tag{1}$$

This general form of $G$ is the solution of a set of simultaneous equations describing a steady state of successive depositions of units at rate $A$ and their detachment at rate $B$. The first factor in Eq. 1 is the "barrier" factor, determined by the barrier to deposition of the unit, while the term in brackets is the "driving force", or "survival" factor. In the coarse-grain secondary nucleation theory of Lauritzen and Hoffman (LH) [16, 17], the barrier factor is the rate $A_0$ of deposition of the first stem in a new layer of stems on the growth face (Fig. 5). According to LH the main barrier is the excess side- surface free energy $2bl\sigma$ for the creation of the two new surfaces either side of the single-stem nucleus. Thus

$$A_0 = \beta \, e^{-\frac{2bl\sigma}{kT}} \, e^{-\frac{2b^2\sigma_e}{kT}}$$

where $\beta$ is the pre-exponential factor, $\sigma$ and $\sigma_e$ are the side- and end- (fold-) surface free energies and $b$ is the width of a chain. For subsequent stems in the surface patch

$$A = \beta \, e^{-\frac{2b^2\sigma_e}{kT}}$$

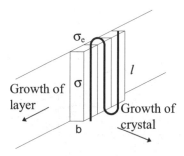

**Fig. 5** Chain deposition on the side surface of a polymer crystal. $\sigma$ and $\sigma_e$ are side-surface and end- (fold-) surface free energies, and $b$ is the width of the chain (after [16])

The detachment rate of a stem is

$$B = \beta\, e^{-\frac{2b^2\Delta\phi}{kT}}$$

where $\Delta\phi$ is the bulk free energy of crystallization, equal to $\Delta T \Delta h_f / T_m$ ($\Delta h_f$ is the heat of fusion, and $\Delta T = T_m^0 - T_c$ the supercooling, with $T_m^0$ the equilibrium melting temperature).

Thus from Eq. 1,

$$G = \beta\, e^{-\frac{2blo}{kT}}\, e^{-\frac{2b^2\sigma_e}{kT}} \left(1 - e^{-\frac{2b^2 l \Delta h_f \Delta T}{kTT_m}}\right) \tag{2}$$

The barrier factor, i.e. the expression before the bracket, decreases exponentially with increasing stem length $l$ and the driving force factor (the term in brackets) increases with $l$ (see Fig. 6). The product $G$ has a maximum at an $l$ value which is considered as the kinetically determined lamellar crystal thickness in polydisperse polymers. $G$ becomes positive at the minimum stable fold length $l_{min}$:

$$l_{min} = \frac{2\sigma_e T_m}{\Delta h_f \Delta T} \tag{3}$$

Since in monodisperse oligomers $l$ adopts only discrete values $L/m$, the $G(l)$ curve is sampled only at these allowed stem lengths, as illustrated schematically in Fig. 6 for $l = L$ and $l = L/2$. For modest undercoolings, changing the crystallization temperature affects only the driving force factor, as shown in Fig. 6 for three temperatures $T' > T'' > T'''$. At $T'$ and $T''$, $l_{min} > L/2$, hence only extended-chain crystals can grow. However, once $l_{min} < L/2$, the growth rate of once-folded chain $F_2$ crystals becomes positive and, due to the higher attachment rate $A_0$, it increases steeply with increasing $\Delta T$ to overtake the growth rate of extended chains.

Studies on PEO fractions in the molecular weight range between 1500 and 12 000 have shown [18] that the slope of the increasing crystal growth rate with supercooling $dG/d(\Delta T)$ does indeed increase abruptly at a series of spe-

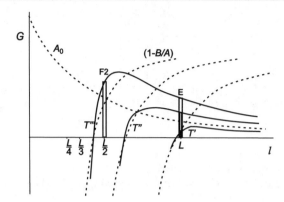

**Fig. 6** The "barrier" factor, which is equal to $A_0$ in LH theory, and the "driving force", or "survival" factor $(1 - B/A)$ of the growth rate as functions of stem length (schematic). $A_0$, $A$ and $B$ are rates of attachment of the first and subsequent stems, and rate of detachment, respectively. The $(1 - B/A)$ factor is drawn for three temperatures $T' > T'' > T'''$. *Vertical rectangles* show growth rates for discrete integer folded forms E and $F_2$ of a monodisperse oligomer

cific crystallization temperatures $T_c^{F_x-F_y}$. These mark the transitions between the growth of E and $F_2$ crystals, or between $F_2$ and $F_3$ crystals, etc. Figure 7 shows the temperature dependence of the growth rate from the melt normal to the {010} lateral face for five sharp low molecular weight fractions and for an unfractionated polymer (note the log rate scale). The breaks in the slope at $T_c^{F_x-F_y}$ are particularly pronounced in low molecular weight fractions and are completely lost in the polymer. As discussed above with reference to Fig. 6, the existence of such sharp increases in $dG/d(\Delta T)$ is, at least qualitatively, entirely predictable by the secondary nucleation theory once quantization of lamellar thickness is recognized. Considering the dilution of folds by chain ends, the supercoolings at which growth transitions occur, relative to the melting point of extended chains, are given by [19]

$$\Delta T_c^m = \frac{2\sigma_{e(\infty)}T_m}{\Delta h_f}\left(\frac{m-1}{m}\right) \bigg/ \left[\left(\frac{L}{m}\right) - \delta\right] \tag{4}$$

Here, $\sigma_{e(\infty)}$ is the fold-surface free energy for $L = \infty$ and $\delta$ is a small constant, reflecting the fact that some finite supercooling is required for crystal growth. In the case of $m > 1$, $\delta$ also accounts for the additional supercooling necessary for $G(F_m)$ to overtake $G(F_{m-1})$ (see Fig. 6).

For polymer crystals of fixed fold length $l$, or for alkane crystals in a given $F_m$ form, the slope $dG/d(\Delta T)$ should be constant within the limited temperature interval that is accessible for measurement [20]. The expansion of the supercooling-dependent factor in brackets in Eq. 2 gives the above-mentioned linear relationship on supercooling

$$G = K \left( T_m^{F_m} - T \right)$$

where

$$K = \frac{b^2 l \Delta h_f}{kTT_m} \beta e^{-\frac{jblo}{kT}} e^{-\frac{2b^2 \sigma_e}{kT}} \tag{5}$$

According to the LH theory, depending on whether mononucleation (Regime I) or polynucleation (Regime II) is involved, $j$ is equal to 2 or 1, respectively [20].

Indeed, when plotted on a linear scale, the growth rate $G$ vs. $\Delta T$ dependence in Fig. 7 does not deviate much from linearity for a given value of $m$ [21]. However, linear $G$ vs. $\Delta T$ is also consistent with the alternative roughness-pinning polymer crystallization theory of Sadler [22]. According to this theory, the difference between nucleation and growth of molecular layers on the growth face is negligible and the growth surface is rough rather than smooth.

On closer scrutiny the crystal growth data for PEO fractions deviate somewhat from those expected from secondary nucleation theory; there were departures from linearity of $G$ vs. $\Delta T$ just above the $T_c^{F_x - F_y}$ transition temperatures [21, 23–25] (see also Sect. 4.3.4) and the value of $\sigma$ derived from the kinetics was therefore inconsistent, reaching unexpectedly low values of around $2 \, \mathrm{mJ/m^2}$, as compared to an accepted value around $10 \, \mathrm{mJ/m^2}$ [23]. Kovacs and Point thus argued that the LH theory had serious flaws.

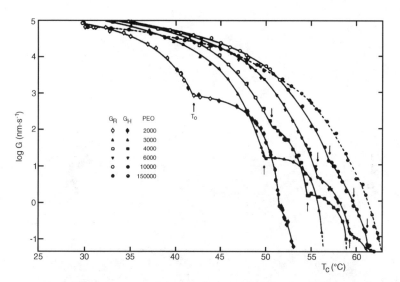

**Fig. 7** Temperature dependence of growth rate $G_H$ of {010} faces of single crystals and $G_R$ of spherulites for five low molecular weight fractions and a polymer of PEO (from [18] by permission of John Wiley & Sons)

# 3
# Manifestations of Self-Poisoning

## 3.1
## Minima in Temperature Dependence of Growth Rate

### 3.1.1
### Self-Poisoning in Melt and Solution Crystallization

Crystallization rate experiments on monodisperse $n$-alkanes have shown a picture rather different from that in PEO. Not only is there a sharp increase in $dG/d(\Delta T)$ at the transition between extended and once-folded (or NIF) crystallization mode ($T_c^{E-F_2}$), but the $dG/d(\Delta T)$ gradient actually becomes *negative* as $T^{E-F_2}$ is approached from above [26]. This was found to occur both in crystallization from melt [26] and from solution [27], and it applies both to crystal growth and primary nucleation [28]. In the early studies [26, 27] only the evolution of the bulk crystalline fraction was monitored. More recently growth rates have been measured directly using phase and interference contrast microscopy of both melt and solution crystallization. Due to the high growth rates of alkane crystals, accurate measurements over a reasonable range of $\Delta T$ values was made possible only through the use of low-inertia temperature-jump cells [28, 29]. Figure 8 shows the growth rate

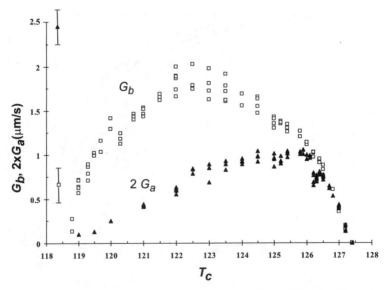

**Fig. 8** Isothermal crystal growth rates along [010] ($G_b$) and [100] ($G_a$) for alkane $C_{210}H_{422}$ from the melt as a function of temperature [30]

dependence on temperature in two perpendicular directions, along [100] ($G_a$) and along [010] ($G_b$), for single crystals of $n$-alkane $C_{210}H_{422}$ grown from the melt [30]. The figure covers the entire extended-chain range. At small super-cooling both rates increase with $\Delta T$, but then they reach a maximum. On further decrease in $T_c$ the rates decrease to a very low value at the transition to folded-chain growth $T_c^{E-F_2}$. At still lower $T_c$ the rates increase steeply (indicated by arrows) as crystallization in the NIF takes over with $l = L/2$. NIF production in melt crystallization below $T_c^{E-F_2}$ has been confirmed by real-time SAXS [26, 31].

Figure 9 is an example of similar growth rate behaviour in solution crystallization, showing $G_{110}$ (here $G_{110} = G_b \cos \gamma$), see Fig. 10) and $G_{100}(= G_a)$ for $C_{246}H_{494}$ in $n$-octacosane [29]. The figure also shows typical crystal habits observed at selected crystallization temperatures.

To help the reader unfamiliar with polyethylene single crystal habits, Fig. 10 summarizes the most typical types [32]: (a) the rhombic lozenge, normally observed in PE solution crystallized at lower temperatures; (b) the

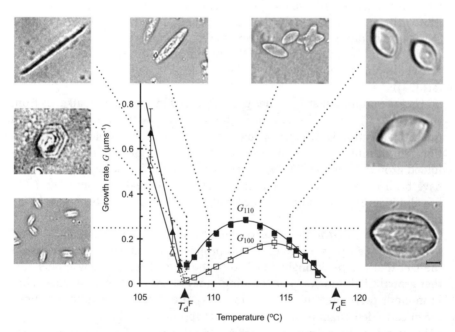

**Fig. 9** Initial rates of growth of $C_{246}H_{494}$ crystals normal to {110} and {100} planes, $G_{110}$ and $G_{100}$, versus crystallization temperature from an initially 4.75% solution in octacosane. The associated interference optical micrographs show typical crystal habits recorded in situ during growth from solution at selected temperatures. The experimental $G$ values are averages over many crystals (from [29] by permission of American Chemical Society)

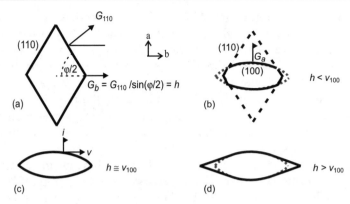

**Fig. 10** Four common types of crystal habit in polyethylene and long alkanes: (**a**) rhombic lozenge bounded by {110} facets; (**b**) lozenge truncated by curved {100} faces (Toda's type B); (**c**) leaf-shaped crystal bounded solely by curved {100} faces (step propagation rate $v$ equals $h = G_{110} / \sin(\varphi/2)$); (**d**) lenticular crystal (Toda's type A) bounded partly by curved {100} and partly by non-crystallographic faceted tangents ($h > v$)

truncated lozenge with curved {100} faces, observed in crystals grown from PE solution at high temperatures and from melt (Type B [33]); (c) leaf-shaped crystals bounded solely by the two curved {100} faces; and (d) lenticular crystals with {100} faces curved in the middle and with straight tangents converging to pointed ends, observed often in melt-crystallized PE (Type A crystals [33]).

The rate minimum at $T_c^{E-F_2}$ has been observed in a number of alkanes from $C_{162}H_{326}$ to $C_{294}H_{590}$ [34–37]. In alkane $C_{294}H_{590}$, where the crystallization rate from solution is sufficiently low for $T_c^{F_2-F_3}$ to be accessible, a crystallization rate minimum has been observed between the once-folded and twice-folded growth intervals [38, 39]. A series of minima, including that at $T_c^{F_3-F_4}$, have been observed in $C_{390}H_{782}$ (see Sect. 5.2). A weak minimum at $T_c^{E-F_2}$ has also been reported in the melt crystallization of a methyl-terminated PEO fraction [40].

The negative $dG/d(\Delta T)$ observed in alkanes has been attributed to an effect termed "self-poisoning" [26]. This proposed mechanism takes account of the fact that chains wrongly attached to the crystal surface may hinder further growth. This blocking, or "pinning", may be overlooked in polymers, but in monodisperse oligomers it becomes very pronounced when, for example, the once-folded-chain melting point $T_m^{F_2}$ is approached from above. Although only extended-chain depositions are stable, close to $T_m^{F_2}$ the lifetime of folded-chain depositions becomes significant. Since extended chains cannot grow on a folded-chain substrate [18], growth is temporarily blocked until the folded-chain overgrowth detaches. At $T_m^{F_2}$ itself, the detachment rate of folded chains drops to the level of the attachment rate and the entire surface is blocked for productive extended-chain growth. The self-poisoning mechanism will be elaborated further below.

### 3.1.2
### Extraneous Poisoning Minima in Binary Alkanes

Recently, a new type of growth rate minimum has been induced above $T_c^{E-F_2}$ in the melt crystallization of several $n$-alkanes by the admixture of a shorter alkane with a melting point between $T_m^E$ and $T_m^{F_2}$ of the host [7, 31]. Figure 11 illustrates this by using the example of a 75 : 25 (w/w) mixture of $C_{162}H_{326}$ and $C_{98}H_{198}$. As seen, there is a general retardation in growth compared to crystallization of pure alkane $C_{162}H_{326}$, a fact also reported by Hosier et al. [41]. However, here a relatively small but reproducible minimum is also observed at 116 °C, in the middle of the extended-chain growth range of $C_{162}H_{326}$; this temperature coincides with the melting point of $C_{98}H_{198}$. The negative $dG/d(\Delta T)$ slope above 116 °C is in this case not due to self-poisoning, but rather to poisoning by the $C_{98}H_{198}$ chain attachments which are nearly stable just above the $T_m$ of $C_{98}H_{198}$ and are thus blocking the growth face of the host crystals. Below the temperature of the minimum there is no drastic increase in $dG/d(\Delta T)$ since (a) $C_{98}H_{198}$ cannot grow rapidly as a minority component, and (b) the growth is retarded by folded-chain $C_{162}H_{326}$ depositions due to the proximity of $T_c^{E-F_2}$ of $C_{162}H_{326}$. Indeed, below $T_c^{E-F_2} = 112.5$ °C there is a very steep increase in $G$ as chain-folded crystallization of $C_{162}H_{326}$ takes

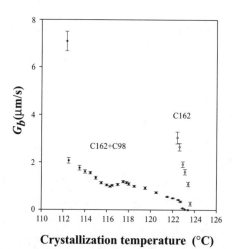

**Fig. 11** Crystal growth rate $G_b$ vs. temperature for a 75 : 25 w/w mixture of $C_{162}H_{326}$ and $C_{98}H_{198}$. The "poisoning" minimum at the melting point of $C_{98}H_{198}$ (116 °C) is seen. The steep increase in $G$ below 112.5 °C (out of scale) is due to the transition to once-folded $C_{162}H_{326}$ growth (from [31])

over, incorporating the guest $C_{98}H_{198}$ chains in the semicrystalline NIF-type structure described in [42] and [43].

Similar poisoning minima are also observed in some other binary alkane mixtures [30].

## 3.2
## Anomalies in Concentration Dependence of Growth Rate

### 3.2.1
### Negative-Order Growth Kinetics and Autocatalytic Crystallization

Another manifestation of self-poisoning is the anomalous negative- order kinetics of crystallization from solution. In long alkanes there is a supersaturation range in which the crystal growth rate actually *decreases* with increasing concentration $c$ [44, 45]. For $n$-$C_{246}H_{494}$ this range is from 0.5 to ca. 2.3% at the particular crystallization temperature of the experiment in Fig. 12 [45]. Both at lower and at higher concentrations the slope $dG/dc$ is positive. The position of the minimum on the concentration scale shifts as $T_c$ is changed. To the left of the minimum extended-chain crystals grow, and to the right of the minimum the growth is in the once-folded $F_2$ form. In the range of negative $dG/dc$ slope, the reaction order drops to as low as – 5 to – 6. Similar behaviour has been reported for $C_{198}H_{398}$ crystallization from phenyldecane [44], and is also observed in other $n$-alkanes.

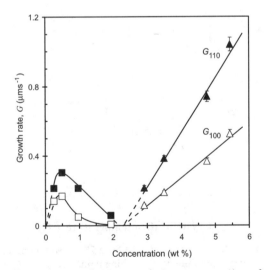

**Fig. 12** Initial growth rates $G_{110}$ and $G_{100}$ vs. solution concentration of $C_{246}H_{494}$ in octacosane at 106.3 °C. *Solid symbols*: extended chain; *open symbols*: once-folded chain

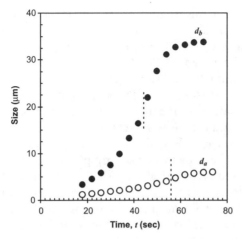

**Fig. 13** Size vs. crystallization time for a $C_{246}H_{494}$ crystal growing from an initially 4.75% solution in octacosane at 108 °C. $d_a$ and $d_b$ are crystal dimensions along the [100] (width) and [010] (length) directions

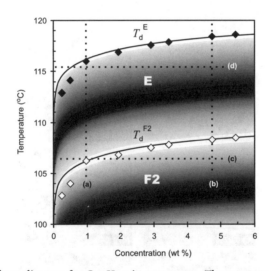

**Fig. 14** Binary phase diagram for $C_{246}H_{494}$ in octacosane. The top curve shows the equilibrium liquidus for extended-chain crystals, and the bottom line the metastable liquidus for once-folded crystals. Experimental dissolution temperatures are fitted to the Flory–Huggins equation with $\chi = 0.15$ (*solid lines*). *Vertical dotted lines* (a) and (b) indicate the concentrations at which the growth rates were determined as a function of $T_c$ in [29]. *Horizontal dotted lines* indicate the temperatures at which the rates were determined in [45] as a function of concentration. $G(c)$ at $T_c = 106.3$ °C, measured along line (c), is shown in Fig. 12. The *shading* indicates schematically the crystal growth rate (*black* = fast), and the *dashed line* the position of the growth rate minimum

One consequence of the negative-order kinetics is autocatalysis of the crystallization process. If the initial concentration is higher than that of the rate maximum, crystal growth will accelerate initially as the concentration decreases. This is illustrated in Fig. 13. Once past the concentration of the growth-rate maximum, the rate drops off. Interestingly, the positions of steepest slope in the time dependencies of crystal length and width do not coincide (Fig. 13), as the positions of the maxima in $G_{110}$ and $G_{100}$ differ (Fig. 12).

The $G(c)$ minimum can be easily related to the $G(T)$ minimum if one considers the alkane–solvent phase diagram. For $C_{246}H_{494}$ in octacosane this is shown in Fig. 14. The growth rate at any particular position in the phase diagram is indicated schematically by shading, with black indicating high and white low rate. It is easy to see how, at certain temperatures, the growth rate would pass through a maximum and a minimum as a function of either temperature (vertical cut) or concentration (horizontal cut). The $G(c)$ dependence in Fig. 12 corresponds to the horizontal line (c) in Fig. 14.

The changes in crystal habit with changing concentration of crystallizing solution were found to be as profound as the changes with crystallization temperature. As seen in Fig. 15, the hexagonal truncated lozenge crystal at low $c$ (0.25%) changes to leaf-shaped and then needle-like crystals at the rate minimum (cf Fig. 12). Thereafter, as folded-chain growth sets in (Fig. 15e), an

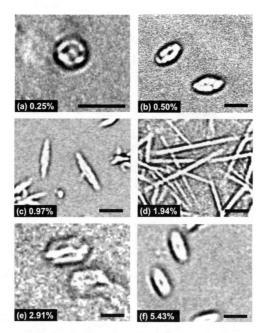

**Fig. 15** Interference optical micrographs of $C_{246}H_{494}$ crystals in octacosane at $T_c =$ 106.3 °C at increasing concentrations of initial solution. The *bar* represents 10 μm

abrupt change to the faceted truncated lozenge takes place, turning to a leaf shape at higher concentration as the cycle repeats itself.

### 3.2.2
### "Dilution Wave"

If the initial solution concentration $c$ is somewhat higher than that of the minimum in $G$, small folded-chain crystals will form and soon stop growing as the metastable equilibrium with the solution (lower liquidus in Fig. 14) is reached. Figure 16a shows such folded-chain crystals in the case of $C_{198}H_{398}$ grown from phenyldecane. What then follows is a phenomenon described as a "dilution wave" [44]. As both nucleation and growth of extended-chain crystals are highly suppressed under these conditions (growth rate minimum), no visible change occurs for some considerable time, after which, rather

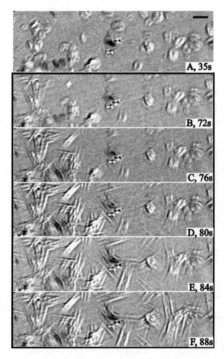

**Fig. 16** Series of interference contrast optical micrographs of an initially 4.2 wt % solution of $n$-$C_{198}H_{398}$ in phenyldecane at successive times (indicated) upon reaching $T_c = 97.4\,°C$. The progress of the "dilution wave" is shown in (**b**) through (**f**), triggering the processes of crystallization of needle-like extended-chain crystals and simultaneous dissolution of folded-chain crystals. The needles form along the two {100} faces of the truncated lozenge shaped folded-chain crystals, with a third parallel crystal often appearing in the middle. *Bar* = 20 μm. (From [44] by permission of American Physical Society)

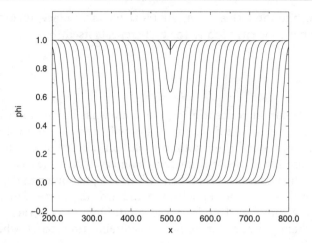

**Fig. 17** Simulation of two dilution waves initiated by a small, localized perturbation. There is a fixed time period between each of the curves. The parameter $\phi$ is the scaled difference between concentrations of saturated solutions $c^E$ and $c^{F_2}$ in equilibrium with, respectively, extended and folded-chain crystals at temperature $T_c$, such that $c = c^E + (c^{F_2} - c^E)\phi$ (from [46] by permission of American Institute of Physics)

suddenly, the folded-chain crystals are replaced by needle-shaped extended-chain crystals. The transformation sweeps through the suspension (from left to right in Fig. 16b–f). As the first extended-chain crystal forms successfully, it depletes the surrounding solution and thus triggers the growth of other extended-chain crystals in the vicinity. A dilution wave is thus generated, easing the inhibitory effect of high concentration on extended-chain growth.

The reaction diffusion equation describing the dilution wave has been found [46] to coincide with that used in genetics to describe the spread of an advantageous gene [47]. The wave spreading rate is determined by the polymer diffusion constant in the solution and the crystal growth rate $G^E$. Figure 17 presents a series of solutions of this equation for successive times, showing the initial dip in concentration spreading in both directions.

# 4
# Events at the Growth Surface

## 4.1
## A Simple Theory of Self-Poisoning

The growth rate minima cannot be explained by the secondary nucleation theory in its present form. The two major simplifications that the LH theory makes are: (a) neglect of the segmental ("fine-grain") nature of stem deposition, and (b) neglect of competing attachments of stems with different, particularly shorter, length. On the other hand, the roughness-pinning theory of Sadler [22], which does not suffer from these simplifications, allows for the occurrence of growth rate minima. In an approximate rate-equation treatment of a system with certain preferred lamellar thicknesses, Sadler and Gilmer observed relatively shallow minima as a function of supercooling [48], but were reluctant to publish until the first minimum was observed experimentally in alkanes [26]. The observed deep minimum in $G$ was reproduced more quantitatively when the one-dimensional model was simplified so that it could be solved exactly [49]. The extended chain was split into two half-chains and thus it would attach in two stages. Alternatively, if the second half-chain attaches alongside the first half-stem rather than as its extension, a folded chain results, carrying the extra free energy of folding (see Fig. 18) [49]. The forbidden process is the attachment of an extended chain onto a folded chain. A Monte Carlo simulation study, where these same selection rules were applied, has also reproduced the kinetics of the self-poisoning minimum and, since the model was two-dimensional, it provided some additional insight into the crystal shape [49].

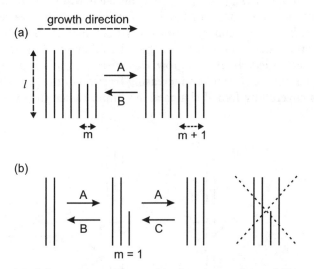

**Fig. 18** The model of elementary steps as used in the rate equation and Monte Carlo simulation treatments that reproduced the self-poisoning minimum. A cross-section (row of stems) normal to the growth face is shown. There are three elementary steps differing in their barrier and driving force: attachment (rate A) and detachment (rate B) of segments equal to half the chain length, and partial detachment of an extended chain (rate C). The key self-poisoning condition is that attachment of the second half of an extended chain is allowed only if $m = 1$, i.e. an extended chain cannot deposit onto a folded chain (from [49] by permission of the American Institute of Physics)

The rate equation model of [49] has subsequently been simplified further and adapted to include solution concentration dependence [44] (see Fig. 19). Instead of depositing in two stages, here the attachments of an extended chain as well as of a folded chain are one-step processes. For extra simplicity, it was assumed that the rates of the two processes are equal ($= A$). However, the detachment rates $B^E$ and $B^{F_2}$ differ. As before, the critical condition that an extended chain cannot attach on top of a folded chain was kept. This ensured that the minimum in $G(T)$ is reproduced even by this "coarse-grain" model. Thus it can be said that, of the two major simplifications characterizing the LH theory, it is the neglect of competing stem attachments that fails it primarily, rather than neglect of the segmental nature of stems.

The concentration dependence of $G$ was treated by making $A$ a linear function of $c$ and adding the mixing entropy to the free energies $\Delta F_0^E$ and $\Delta F_0^{F_2}$ of the pure alkane:

$$\Delta F^E = \Delta F_0^E + kT \ln c \tag{6a}$$

$$\Delta F^{F_2} = \Delta F_0^{F_2} + kT \ln c \tag{6b}$$

The retardation in crystal growth with increasing concentration, calculated according to this model, is shown in Fig. 20. The parameters used apply to crystallization of $C_{198}H_{398}$ (see [44]), but the calculated growth curves match qualitatively those measured for $C_{246}H_{494}$ (see Fig. 12). This effect can be explained as follows. For concentrations less than that of the rate minimum ($c_{min}$), the attachment rate $A$ is lower than the folded-chain detachment rate $B^{F_2}$, so only extended chains can grow. However, as $c_{min}$ is approached $A$ becomes comparable to $B^{F_2}$ and an increasing fraction of the growth face equal to $A/B^{F_2}$ is covered by folded chains at any time. Hence the extended-chain

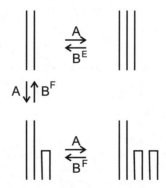

**Fig. 19** Row model as in Fig. 18 but simplified further. This "coarse-grain" model assumes a one-step attachment and detachment of a whole chain, either in extended or folded conformation. Attachment rate $A$ is assumed to be the same in both cases, while detachment rates $B^E$ and $B^{F_2}$ differ (from [44] by permission of American Physical Society)

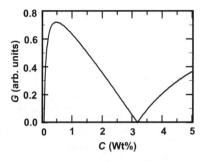

**Fig. 20** Crystal growth rate $G$ of $n\text{-}C_{198}H_{398}$ calculated as a function of concentration at constant $T_c$ using the model in Fig. 19. Parameters are those for $C_{198}H_{398}$ crystallization from phenyldecane at $T_c = 98.0\,°C$, experimentally measured in [44]. Compare with Fig. 12 (from [44] by permission of American Physical Society)

crystal growth rate is

$$G^E = (1 - A/B^{F_2})(A - B^E) \tag{7}$$

and as $c \to c_{min}$ and $A \to B^{F_2}$, so $G^E \to 0$.

An additional slow process of chain extension underneath the obstructing overgrowth has also been considered in [49]. The second half of the chain was allowed to attach to the half-chain nearest to the extended-chain substrate at rate $D$. This process, mimicking the "sliding diffusion", preserved a finite growth rate even at the rate minimum, i.e. it made the minimum shallower.

### 4.2
### Growth on {100} and {110} Faces: Comparison with Polyethylene

The cause of the changes in crystal habit with changing $T_c$ in polyethylene, or in any other polymer, is unknown. These changes have a profound effect on the morphology of industrial polymers. Thus in melt-solidified polyethylene, crystals grow along the crystallographic $b$-axis with lenticular morphology similar to that formed from polyethylene solutions at the *highest* $T_c$ [50, 51] (Fig. 10b–d), or else from long alkanes close to the *lowest* $T_c$ within the extended-chain regime (Figs. 9 and 15) or any folded-chain regime ($F_2$, $F_3$ ...). In the absence of twinning or branching, the lateral habits in polyethylene and long alkane crystals are determined by only two parameters: (a) the ratio of growth rates normal to $\{110\}$ and $\{100\}$ faces, $G_{110}/G_{100}$, and (b) the ratio $i/v$ of step initiation and propagation rates on $\{100\}$ faces. We will deal with the former first, and with the latter in the next section.

Figures 9 and 15 show the conspicuous changes in crystal habit occurring within rather narrow intervals of crystallization temperature and concentration, particularly around the self-poisoning minimum. Similar diversity of shape is observed in melt crystallization [30]. In addition to the morphologies seen in $C_{246}H_{494}$, in shorter alkanes such as $C_{198}H_{398}$ and $C_{162}H_{326}$ rhombic lozenges, bounded solely by {110} facets, are observed at the highest temperatures [52]. Figure 21 shows typical crystals grown at the same supercooling of $\Delta T = 3\,°C$ from three different alkanes.

Thus, in solution-grown alkanes the general morphological cycle starts with lozenge crystals, changing to truncated lozenge with the relative length of {100} faces increasing and {110} receding as $T_c$ is lowered or as $c$ is increased (cf Fig. 10). This means that the ratio of growth rates $G_{100}/G_{110}$ decreases with increasing supercooling or supersaturation, until the rate minimum is reached. This is illustrated in Fig. 22 for $C_{198}H_{398}$, and in [29] for

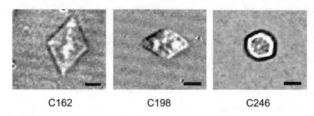

C162                    C198                    C246

**Fig. 21** Micrographs of extended-chain crystals of the alkanes indicated grown from 1 wt % octacosane solutions at the supercooling temperature of 3.0 °C

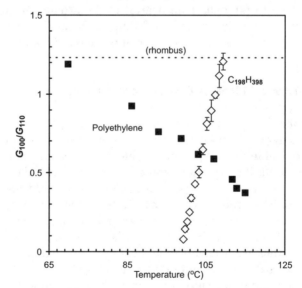

**Fig. 22** Ratio of growth rates $G_{100}/G_{110}$ vs. crystallization temperature for extended-chain crystals of $C_{198}H_{398}$ from 2% (w/v) solution in octacosane (*diamonds*) and for linear polyethylene from 0.05% solutions in hexane (*squares*). The polyethylene crystal at $T_c = 70.0\,°C$ and the $C_{198}H_{398}$ crystals above $T_c = 110\,°C$ are non-truncated lozenges; these form for any $G_{110}/\cos(\varphi/2) \geq G_{100}$, where $\varphi/2 = \tan^{-1}(a_0/b_0)$ and $a_0, b_0$ are unit cell parameters (cf Fig. 10). Data for PE are from [32] (from [45])

$C_{246}H_{494}$. It is interesting to note that this trend is exactly the opposite of that in polyethylene, where the $G_{100}/G_{110}$ ratio decreases with *increasing* $T_c$ [32] (see Fig. 22). The extremely low $G_{100}/G_{110}$ ratio of $\leq 1 : 15$, which gives rise to the long needle-like crystals near the rate minimum, is not seen in PE.

The facts that in alkanes $G_{100}/G_{110} \to 0$ at the rate minimum, and that the morphological cycle is repeated at the $F_2$–$F_3$ [39] and $F_3$–$F_4$ minimum [53], suggests strongly that the depression of $G_{100}/G_{110}$ is the result of self-poisoning. One may thus conclude that self-poisoning preferentially retards the growth on $\{100\}$ faces.

It is well known that in PE, rhombic lozenges are only seen at the lowest $T_c$, e.g., below 75 °C in xylene solution. The surprising finding that non-truncated $\{110\}$-bounded lozenges are seen at the *highest* $T_c$ in alkanes is attributed to the *exceptionally weak self-poisoning* well above the folded-chain dissolution temperature $T_d^{F_2}$. It is thus understandable that rhombic crystals should form in shorter alkanes rather than in longer ones, as in the former the temperature difference $T_d^E - T_d^{F_2}$ is larger, hence $T_d^{F_2}$ is more remote and the lifetime of the obstructing folded deposition is short. Such a situation of negligible poisoning cannot occur in polydisperse PE, since at any $T_c$ there are stems somewhat shorter than $l_{min}$ (Eq. 3), with sufficient lifetime to cause obstruction to growth as well as a higher attachment rate because of a lower barrier (self-poisoning or pinning [22]). The fact that in PE the signs of $G_{100}$ obstruction are most pronounced at highest temperatures (lowest $G_{100}/G_{110}$) suggests that self-poisoning increases either with temperature and/or lamellar thickness.

A snapshot of a cut normal to the surface of a growing crystal might thus look as in Fig. 23. This situation corresponds to the "kinetic roughness" in the roughness-pinning model of Sadler [22] (see also [71]). The unstable overgrowth must detach before the stem underneath can rearrange. The reason

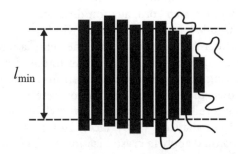

**Fig. 23** A possible instantaneous configuration of a row of stems perpendicular to the crystal surface schematically illustrating self-poisoning in a polydisperse polymer. The same picture could apply to a layer of stems depositing onto the crystal surface and parallel to it; in that case it would illustrate the retarding effect on layer spreading rate $v$

for self-poisoning in polyethylene being more pronounced at higher $T_c$, as suggested by morphology, is one of the open questions.

## 4.3
## Step Initiation and Propagation Rates

### 4.3.1
### Rounded Lateral Crystal Faces

The discovery of rounded polymer crystal edges [52, 54] has motivated Sadler to propose the rough-surface theory of polymer crystal growth [22, 48]. His theory was intended to describe crystallization at higher temperatures, where lateral crystal faces are above their roughening transition temperature $T_R$. High equilibrium roughness removes the necessity of surface nucleation as there is no surplus free energy associated with a step. In a simple cubic crystal the roughening transition on a $\{100\}$ surface should occur at $kT/\varepsilon \approx 0.6$, where $\varepsilon$ is the pairwise nearest-neighbour attraction energy. If whole stems are treated as non-divisible units on a polymer crystal surface, then $\varepsilon$ is so large that the surface at equilibrium is always smooth and flat. However, a simple estimate shows that, if polyethylene stems were broken down into segments of $\leq 6$ $CH_2$ units, the surface would be thermodynamically rough. This crude estimate does not take account of connectivity between segments and there are numerous problems in estimating the $T_R$ of such a complex system. However, Sadler argued that the fact that the $\{100\}$ faces in polyethylene become curved at high temperatures indicates the presence of the roughening transition. In contrast, $\{110\}$ faces with somewhat denser packing of surface chains and hence a higher $\varepsilon$ would have a higher $T_R$, hence the $\{110\}$ faces remain faceted.

However, subsequent detailed studies [33, 55, 56] have shown that the curvature of $\{100\}$ faces in polyethylene can be explained quantitatively by applying Frank's model of initiation and movement of steps [57]. Curvature occurs when the average step- propagation distance is no more than about two orders of magnitude larger than the stem width.

Hoffman and Miller [58] proposed that the retardation in $v$, leading to curved crystal edges, comes from lattice strain imposed by the bulky chain folds in the depositing chains; this results in strain surface energy $\sigma_s$ which acts by slowing down the substrate completion rate $v$. In order to be consistent with the observed morphology in polyethylene, $\sigma_s$ would have to be higher on $\{100\}$ faces than on $\{110\}$ faces. However, it is difficult to see how this argument would apply to crystallization of extended-chain $n$-alkanes with no chain folds. The evidence from alkanes presented below suggests that the slow step propagation is largely due to self-poisoning. This is supported by the observations on PEO (Sect. 4.3.4).

### 4.3.2
### Suppression of $i$ and $v$ on {100} Faces

In order to determine the rate of initiation ($i$) and propagation ($v$) of steps, the profiles of the curved "{100}" faces in solution- and melt-crystallized alkanes may be fitted to those calculated according to Mansfield [55] and Toda [59], taking account of the centred rectangular lattice representative of polyethylene and alkanes. This was carried out recently in a systematic way in a series of alkanes crystallized both from melt [30] and solution [29, 39, 45, 53, 60]. As an example, the $i$ and $v$ values thus obtained are plotted against $T_c$ for $C_{246}H_{494}$ in Fig. 24 [29]. While both the secondary nucleation rate ($i$) and propagation rate of steps ($v$) pass through a maximum and a minimum as a function of $T_c$, and then increase steeply in the chain-folded temperature region, there is a significant difference in the shape of the two graphs near the rate minimum. In this and other examples of solution crystallization, $i$ is considerably more suppressed by self-poisoning than $v$; while $v$ drops to half its maximum extended-chain value at the minimum, $i$ is reduced to only 1/500 of its value at the maximum. Furthermore, the slope $di/d(\Delta T)$ turns negative higher above $T_c^{E-F_2}$ than the corresponding slope $dv/d(\Delta T)$. The dis-

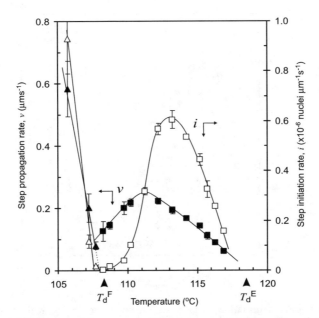

**Fig. 24** Step initiation rate $i$ (*open symbols*) and step propagation rate $v$ (*solid symbols*) on the {100} faces as a function of crystallization temperature for $n$-$C_{246}H_{494}$ crystals growing from a 4.75% octacosane solution. Key: *squares*, extended chain; *triangles*, folded-chain growth (from [29])

proportionately large retardation in $i$ near the minimum causes increasing elongation of the Mansfield ellipse, so that $\{100\}$ faces become virtually flat (see needle-like crystals in Fig. 9).

It is informative to observe the temperature dependence of the initiation/propagation ratio. Figure 25 shows $ib^2/2v$ against crystallization temperature for four long alkanes. This dimensionless parameter gives the ratio against the probability of its attachment to a niche on either side of a spreading layer (for polyethylene lattice $b^2 = a_0b_0/2$). The higher the parameter, the higher the curvature. The $ib^2/2v$ ratio is seen to depend strongly on temperature. Within the extended-chain growth range it passes through a maximum, descending towards both the low- and the high-temperature ends of the range. In the folded-chain growth regime the trend repeats itself, as shown in Fig. 25 for $C_{294}H_{590}$. The maximum value of the ratio is lower for the longer alkanes $C_{294}H_{590}$ [45] and $C_{390}H_{782}$ [53]. This may be the result of the temperature interval of crystallization of a given folded form $F_x$ being narrower for longer chains.

The maximum in $ib^2/2v$ primarily reflects the fact that the maximum in $i$ is narrower than the maximum in $v$. As $T_d$ is approached, $i$ should always tend to zero faster than $v$, due to the barrier for nucleation being larger than that

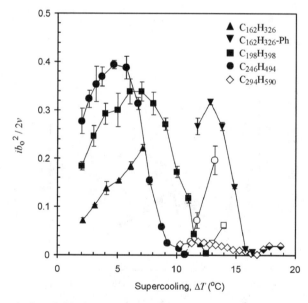

**Fig. 25** Dimensionless initiation to propagation rate ratio $ib^2/2v$ against supercooling $\Delta T = T_d^E - T_c$ for crystals of four $n$-alkanes grown from $n$-octacosane solution (initial concentration $4.8 \pm 0.2$ wt %) and for $C_{162}H_{326}$ from phenyldecane (1.0 wt %). $C_{162}H_{326}$ crystals from phenyldecane (Ph) did not show $\{100\}$ faces at low $\Delta T$. Symbols: *solid* = E, *open* = $F_2$, *half-solid* = $F_3$. Based on data from [39]

for propagation, as long as growth is even marginally nucleation controlled. Ratio $ib^2/2v$ is used in the LH theory to distinguish between growth regimes. According to LH theory, the alkane crystal growth very close to $T_d^{F_x}$ and $T_c^{F_x-F_y}$ would be Regime I, while between these temperatures it would be Regime II or III [16].

Although the minimum in $v$ is not as deep as that in $i$, the mere fact that it exists means that, like step nucleation, step propagation itself is also retarded by self-poisoning. This also implies that surface energy is not the main barrier for stem deposition, as implied by the classical LH theory. Instead, the barrier for step propagation is almost entirely entropic, with self-poisoning, or pinning, being its major component. With reference to Fig. 5, a chain depositing into an existing niche leads to no increase in the side-surface energy, only to an increase in end-surface energy $2b^2\sigma_e$ which, for extended-chain crystals, is very small. $v$ should therefore be particularly fast in extended-chain crystals, with no curvature expected. Obstructing folded-chain depositions would be rare, as their surface energy barrier is significantly higher than that for extended-chain crystals. In fact, contrary to the expectation from LH theory, the curvature of {100} faces is higher in extended-chain than in folded-chain crystals (Fig. 25).

The models used to describe self-poisoning kinetics (Figs. 18 and 19) are based on Sadler's row of stems normal to the growth face ("radial growth"). However, it can be equally applied to step propagation, i.e. "tangential" growth [29, 61]. The two types of rows, to both of which the models in Sect. 4.1 could be applied, are schematically illustrated in Fig. 26.

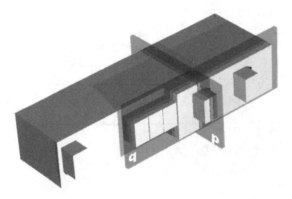

**Fig. 26** Schematic view of the growth face of an extended-chain lamellar crystal poisoned by stems of half the chain length. The row-of-stems model can be applied with the row perpendicular to the growth face, as in the previous "rough growth" models to describe retardation of $i$ (row $p$), or parallel to the growth face to describe retardation of $v$ (row $q$). (From [29], by permission of American Chemical Society)

### 4.3.3
### Growth at Sadler's Rough-Surface Limit

The initiation/propagation ratio $ib^2/2v$ reaches exceptionally high values of $\sim 0.4$ in the middle of the temperature range of extended-chain growth (Fig. 25). Such a high value means that there is almost no preference for a molecule to deposit into an existing niche, i.e. that the growth process virtually ceases to be nucleated. If there was a complete lack of preference for propagation, the ratio would be 0.5. Admittedly, the approximations involved in the derivation of the expression for Mansfield's ellipse [55, 57] are not fully justified in this limiting case, so that the accuracy of the quoted value is limited. In fact, for $ib^2/2v = 0.25$, Mansfield's ellipse on a square lattice becomes a circle, and for $ib^2/2v = 0.5$ the ellipse is elongated perpendicular to the growth face. This discrepancy arises primarily because nucleation on the sides perpendicular to the growth face is ignored, as is step propagation along these sides. Nevertheless, there is no doubt that crystallization in the middle of the extended-chain temperature range is at the very limit of nucleated growth, and that it approaches the rough-surface growth mode of Sadler and Gilmer [22, 62].

Rough-surface growth is equivalent to condensation of vapour into a droplet. Treating the crystal lamella as a two-dimensional droplet, one would expect circular crystals in the limit of completely rough growth. Figure 27 shows an extended-chain crystal of $C_{198}H_{398}$ grown from a methylanthracene solution, which is indeed nearly circular. Circular crystals are also observed in self-poisoned growth of PEO fractions (see Sect. 4.3.4).

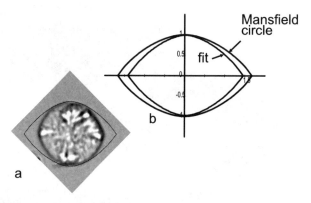

**Fig. 27** (**a**) Nearly circular crystal of $C_{198}H_{398}$ grown from a 2% solution in methylanthracene with Mansfield ellipse for $ib^2/2v = 0.41$ fitted to the {100} faces. Note that here {110} faces are also curved (cf Sect. 4.3.5). (**b**) The Mansfield ellipse, fitted to the crystal in (**a**) compared to the Mansfield circle; both apply to the centred rectangular lattice of polyethylene, rather than to the square lattice

### 4.3.4
### Comparison with Poly(ethylene oxide)

The kinetics of melt crystallization of narrow fractions of PEO has already been mentioned in Sect. 2.2, and the $G(T_c)$ behaviour for one growth face is summarized in Fig. 7. The shapes of melt-grown single crystals have been studied in great detail by Kovacs and co-workers [18, 63, 64]. Particular attention was paid to the narrow temperature range of the extended to folded-chain growth transition. Careful investigation revealed that there are in fact two close transition temperatures, one for each of the two $\{hk0\}$ growth face types. These differences, combined with the thickening process, resulted in the so-called pathological crystals. Provided that the gradients $dG_{hk0}/d(\Delta T)$ are different for different $\{hk0\}$ planes, such differences in $T_c^{E-F_2}$ are inevitable, as $T_c^{E-F_2}$ values are given simply by $G_{hk0}^E = G_{hk0}^{F_2}$. Only where there is a deep minimum at $T_c^{E-F_2}$, as in monodisperse alkanes, do the growth transition temperatures $T_c^{E-F_2}$ for different $hk0$ faces nearly coincide. In fact from Fig. 9 it is clear that $T_c^{E-F_2}$ for $\{110\}$ would occur at a lower temperature than $T_c^{E-F_2}$ for $\{100\}$. However the former cannot be observed because the growth of E crystals stops at $T_c^{E-F_2}\{100\}$. This results in the extreme needle-like crystals (Figs. 9 and 15), which can be regarded as the pathological crystals of $n$-alkanes.

A more general observation in PEO fractions is that more or less faceted crystals grow at most temperatures, except just above the transition temperature $T_c^{E-F_2}$, where crystals are rounded and sometimes indeed circular. This has been confirmed by subsequent studies, as illustrated in Fig. 28 [40].

The increase in curvature of PEO crystal faces immediately above $T_c^{E-F_2}$ (Fig. 28) means a high $ib^2/2\nu$ ratio at that temperature. Thus, compared to alka-

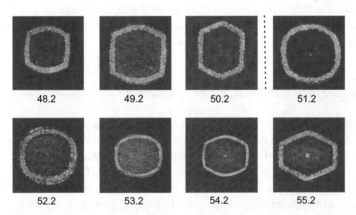

**Fig. 28** Optical micrographs of PEO single crystals (fraction $M_n = 3000$) grown from the melt as a function of crystallization temperature (in °C). The extended–folded chain-growth transition $T_c^{E-F_2}$ is between 51.2 and 50.2 °C (adapted from [40])

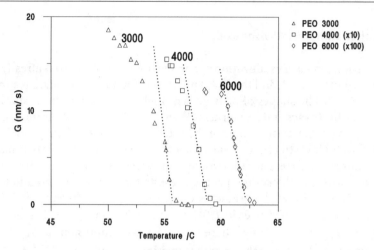

**Fig. 29** Linear growth rates from melt of extended-chain crystals of PEO fractions 3000, 4000 and 6000 molecular weight as a function of crystallization temperature (re-plotted from [18]). *Dashed lines* are maximum slope tangents (after [25])

nes where $ib^2/2v$ passes through a maximum in the middle of a growth branch (E or $F_2$) and drops to a very low value at the $T_c^{E-F_2}$ or $T_c^{F_2-F_3}$ transition (see Fig. 25), in PEO that maximum is at the transition itself. It is interesting that, even when plotted on a linear scale, Kovacs's growth rate data versus $\Delta T$ (Fig. 7) show a conspicuous downward deviation from linearity just above $T_c^{E-F_2}$ (see Fig. 29) [21, 25]. In the past the effect was attributed to impurities [24], but subsequently it was suggested that it is caused by self-poisoning [25]. The conclusion is therefore that, near $T_c^{E-F_2}$, step propagation is preferentially retarded in melt-grown PEO, while $i$ is preferentially retarded in solution-grown alkanes. The reason for this difference is not clear. In melt crystallization of alkanes, $i$ is also preferentially retarded at the rate minimum, although the $ib^2/2v$ ratio does not drop as low as that in solution growth [30].

### 4.3.5
### Curved {110} Faces: Asymmetry in Step Propagation

While extended-chain crystals of $C_{162}H_{326}$ and $C_{198}H_{398}$ grown from octacosane at the smallest supercooling are faceted rhombic lozenges (see Fig. 21a,b), those grown under similar conditions from 1-phenyldecane [52] and methylanthracene [60] have an unusual habit shown in Fig. 30. This habit has not been seen in polyethylene and it has been termed "$a$-axis lenticular" because, unlike the more common lenticular (lens-like) polyethylene crystals, its long axis is parallel to the crystal $a$-axis. In fact the habit can be best described as being bounded by curved {110} faces. The interesting feature is the asymmetry of the

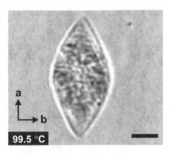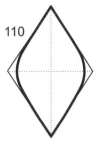

**Fig. 30** *Left*: An "*a*-axis lenticular" crystal of $C_{162}H_{326}$ in 1-phenyldecane grown from an initially 1.0% solution at 99.5 °C. Interference optical micrograph; bar length 10 μm. *Right*: Schematic outline of the crystal, indicating the four {110} sectors (from [52] with permission of the American Chemical Society)

curvature; while the faces are curved at the obtuse apex, they are straight at the acute apex. This has been attributed to the propagation rates of steps on the {110} face being different in the two directions: the rate $v_s$, directed towards the acute apex, is higher than the rate $v_b$, directed towards the obtuse apex [52]. Similar asymmetry may be expected in other polymers where the growth face lacks a bisecting mirror plane normal to the lamella. In fact exactly the same *a*-axis lenticular crystals as in Fig. 30 are seen regularly in α-phase single crystals of poly(vinylidene fluoride) ([65] and [66]; note that *a* and *b* axes are exchanged compared to polyethylene).

## 4.4
### Effect of Solvent: Molecular Pre-ordering in the Liquid

It has been observed that the nature of the solvent has a significant effect on both crystal growth rates and morphology. For example, the growth rates of crystals of $C_{162}H_{326}$ [52] and $C_{198}H_{398}$ [45] from *n*-octacosane solutions were found to be about three times as high as those from solutions in 1-phenyldecane. Furthermore, at low supercooling *a*-axis lenticular extended-chain crystals formed from phenyldecane, while faceted {110} lozenges formed from octacosane (see preceding section). Thus, crystallization processes from phenyldecane are generally slower. In particular, relative to step initiation, step propagation on {110} faces is more retarded in phenyldecane.

A more comprehensive study of this effect has been carried out recently, involving the following aromatic solvents: 2-methylanthracene, *m*-terphenyl, *o*-terphenyl, benzoquinoline and triphenylmethane. Some of these are poor solvents and liquid–liquid phase separation interferes in some cases. However, it is clear that growth rates are slower in all these solvents compared to those from *n*-octacosane. In particular, $G_{100}$ is more retarded than $G_{110}$. Figure 31 shows (a) $G_{110}$ and (b) $G_{100}$ as a function of supercooling in several 2 wt % solutions of $C_{198}H_{398}$.

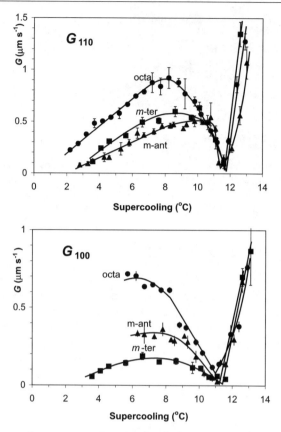

**Fig. 31** Crystal growth rates normal to (**a**) {110} faces and (**b**) {100} faces for alkane $C_{198}H_{398}$ from 2 wt % solutions in *n*-octacosane ($C_{28}H_{58}$) (*circles*), methylanthracene (*triangles*) and *m*-terphenyl (*squares*). Crystals grown from octacosane and methylanthracene at low supercooling do not show {100} faces (from [60])

Compared to octacosane, all other solvents (partly or fully aromatic) used in the study depress the growth rates, particularly $G_{100}$. On the other hand, $G_{110}$ is less affected overall by the solvent type. It is interesting, however, that the minimum in $G_{110}$ is narrower in aromatic solvents. In 2% methylanthracene $G_{110}$ drops very sharply, but only less than 1 °C above the transition to folded-chain growth (Fig. 31); in fact in a 5% methylanthracene solution (not shown), there is no minimum in $G_{110}$ at all.

In aromatic solvents the step propagation rate $v$ seems to be disproportionately more suppressed on both {100} and {110} faces compared to the nucleation rate $i$. This results in *a*-axis lenticular crystals (Fig. 28) and in nearly circular crystals with curvature on both {110} and {100} edges (Fig. 27). Also, as $v$ is suppressed on {100} faces and $G_{110}$ remains relatively high, *b*-axis lenticular crystals with straight-faced pointed ends (type A crystals [33]) are often

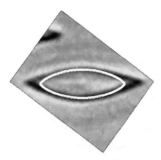

**Fig. 32** Lenticular crystal of $C_{198}H_{398}$ grown from an initially 2 wt % solution in $o$-terphenyl at 115.8 °C. Mansfield ellipse corrected for the alkane crystal lattice is fitted to the central portion of the crystal. Crystal $b$-axis is horizontal [60]

seen (Fig. 32). According to Toda [33], such crystals form when the growth rate in the $b$ direction $G_b = G_{110}/\sin(\phi/2)$ is faster than $v$ on $\{100\}$ faces, hence the spreading layers on $\{100\}$ faces fail to catch up with the spreading substrate.

The detailed study of the solvent effect is still incomplete, but it emerges that the less linear and more "awkward" the shape of the solvent molecule, the more retarded are the crystallization processes. The $\{100\}$ faces, which are generally more susceptible to poisoning, again seem to be more affected by this solvent-induced retardation. We suggest that the effect is due to less pre-ordering of long alkane molecules in aromatic solvents. It is likely that in an $n$-alkane solvent such as octacosane, the long-chain and short-chain molecules form local bundles, with bias towards extended segments, resembling that in the melt [67]. With smaller aromatic solvent molecules such order is less likely. A further comparison between the effects of more linear aromatic molecules (methylanthracene) and less linear molecules ($m$- and $o$-terphenyl, triphenylmethane) suggests that the latter are more effective in retarding the growth process. The growth-promoting effect of linear solvent molecules may thus be interpreted as reducing the conformational entropic barrier for chain extension.

Strobl et al. [68] have recently reported that crystals of poly(ethylene-$co$-octene) grown from $n$-alkane solvents have larger thicknesses than those grown from methylanthracene. This they interpret in terms of the $n$-alkane solution forming a precursor mesophase which enables lamellar thickening by sliding diffusion. However, in crystallization of long alkanes there is no evidence of an equivalent thickening process in octacosane solution.

## 4.5
### Side-Surface Free Energy $\sigma$ from Chain Length Dependence

The lateral surface free energy $\sigma$ is a key parameter in polymer crystallization, and is normally derived from crystallization kinetics. In polydisperse polymers, where the supercooling dependence of growth rate is affected both by changing driving force and lamellar thickness, a product of end-surface and side-surface free energies $\sigma\sigma_\varepsilon$ is obtained. In alkanes, however, Eq. 5 can be used to determine $\sigma$ alone. To avoid ambiguities about the precise values of other parameters in Eq. 5, such as $\sigma_e$, $\beta$ and $\Delta h_f$, $\ln K - \ln l$ can be plotted against $l$ for a series of alkanes and $\sigma$ values determined from the slope of the plot. $K$ was measured close to the melting point of extended chains, above the temperature where $G$ curves downwards due to self-poisoning. Figure 33 shows such plots for extended-chain melt crystallization of the longer alkanes $C_{194}H_{390}$ through $C_{294}H_{590}$ [30]. From the measured rates of growth along the $b$-axis and $a$-axis, values of $\sigma$ on the $\{110\}$ and $\{100\}$ faces, respectively, were obtained, assuming Regime II kinetics ($j = 1$, Eq. 5). The values obtained are $\sigma_{110} = 8.0$ mJ/m$^2$ (or erg/cm$^2$) and $\sigma_{100} = 10.8$ mJ/m$^2$.

These values are in broad agreement with the generally accepted value of around 10 mJ/m$^2$ for $\sigma$ of polyethylene, but provide important differentiation between the two crystallographic faces. The same procedure, when applied to solution crystallization, yielded significantly lower values around 3 mJ/m$^2$, which is attributable possibly to crystal growth being diffusion controlled. For melt crystallization of PEO fractions, Point and Kovacs [23] obtained unusually low values of around 2 mJ/m$^2$ for $\sigma$ determined from the $G$ vs. $\Delta T$ slope close to the $T_c^{E-F_2}$ growth transition. The anomalous value is likely to be due to low apparent $K$ in this region of increasing self-poisoning with increasing $\Delta T$ (see Sect. 4.3.4 and Fig. 29).

It is significant that $\sigma_{100}$ is higher than $\sigma_{110}$. This is consistent with $\{100\}$ faces being more susceptible to self-poisoning and poisoning by impurities, i.e. having a higher entropic barrier for growth. This finding also helps to explain why $\{110\}$-bounded lozenges in polyethylene become increasingly truncated

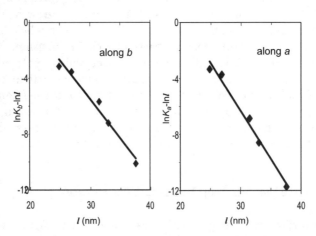

**Fig. 33** Determination of side-surface free energy from chain length dependence of extended-chain crystal growth rates $G_b$ (*left*) and $G_a$ (*right*) from the melt using Eq. 5 [30]

by $\{100\}$ faces at higher crystallization temperatures, where crystal thickness and hence the $2bl\sigma$ barrier is larger.

## 4.6
### Molecularity of Chain Deposition: Two-Dimensional Nucleation

In the concentration region of weak self-poisoning, the concentration dependence of crystal growth rate can provide further insight into the mechanism of the growth process, just as determining the order of a chemical reaction supplies evidence of molecularity of the reaction, or the number of molecules needed to collide to effect a reaction. Since experimental data are now available on concentration dependence of $i$ and $v$ on $\{100\}$ faces for alkane $C_{246} H_{494}$, these were analysed to determine the order $n$ of the attachment rate $A = ca^n$ [69]. At low supercooling $i \approx A - B$ (see Eq. 7), and $B$ is defined through Eq. 6 as $B = (1/c)d$, where $d$ is the detachment rate in pure melt ($c = 1$) at $T_c$. $i$ is therefore given by

$$i = ac^n - (1/c)d \tag{8}$$

An equivalent expression holds for $v$. Equation 8 and its equivalent were fitted to experimental values of $i$ and $v$. The clearest and most striking result is that in the once-folded ($F_2$) crystallization regime, the reaction order for secondary nucleation is found to be 2.5, and for step propagation it is 2.0. Taken literally, this means that 2.5 molecules (five stems) of folded $C_{246} H_{494}$ are involved in the formation of an average critical secondary nucleus, and that deposition at an existing step takes place in units of two molecules (four stems). These values apply to the region of low supersaturation and low $ib^2/2v$. According to the classical nucleation theory, only one stem makes a critical secondary nucleus. The above result, however, supports the idea of a two-dimensional secondary nucleus proposed by Point [70], and by the more recent simulation work of Doye and Frenkel [71]. 2D secondary nucleation has also been postulated by Hikosaka, but only for the highly mobile columnar hexagonal phase [72]. Step propagation by attachment of small clusters of stems has not been discussed specifically, but it is not inconsistent with some simulation studies [73].

## 5
### Roughness of Basal Surfaces

## 5.1
### Upward Fluctuations in Lamellar Thickness

The preceding sections on self-poisoning illustrate the major difficulty of a stable longer stem depositing on top of a shorter stem. The rate minima described above could all be explained by the inability of an extended chain (E) depositing onto a once-folded chain ($F_2$), or an $F_2$ chain depositing onto an $F_3$ chain

(Sect. 4.1). In polymers, lamellar thickness is kinetically determined by the restricted upward fluctuation in stem length on the one hand and, on the other hand, by the high detachment rate of shorter stems with $l \rightarrow l_{min} = 2\sigma_e/(\Delta\phi)$ (downward fluctuation). This problem has been studied analytically [62, 74] and by simulation [71, 75].

The dilution wave effect, described in Sect. 3.2.2, illustrates further the additional impediment to large upward fluctuation caused by self-poisoning. In the experiment described in Fig. 16 it took more than a minute for the first extended-chain nucleus to appear anywhere in the suspension of $F_2$ crystals; however, once the concentration dropped, E crystals nucleated and grew rapidly. The fact that in the mobile hexagonal phase polymers like polyethylene chains can extend behind the growth surface leads to polymer crystals of large thicknesses and with the characteristic wedge-shaped profile [72]. However, in true crystals with 3D order, thickening occurs only close to the melting point, if at all (Fig. 2).

## 5.2
### Rate Minima in Multiply Folded Chains

In the present context it is informative to look at the crystallization kinetics of the longest monodisperse alkane, $C_{390}H_{782}$. Figure 34 shows the growth rate $G_{110}$ from octacosane solution, covering the temperature range of three folded-chain forms: $F_2$, $F_3$ and $F_4$ [53]. The data for $G_{100}$ are qualitatively similar to those for $G_{110}$. While the E-$F_2$ and $F_2$-$F_3$ rate minima are deep, so that growth virtu-

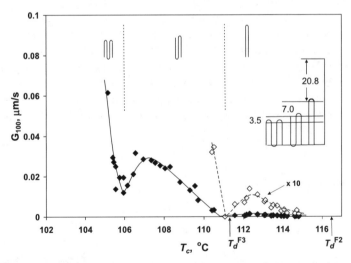

**Fig. 34** Crystal growth rate $G_{110}$ as a function of temperature for $C_{390}H_{782}$ from 4.6 wt % octacosane solution. The temperature ranges of $F_2$, $F_3$ and $F_4$ (part) folded forms is covered. *Open symbols* show values scaled by a factor of 10 and 0.1, respectively. For explanation of the inset, see text

ally stops at the transition temperatures, the $F_3$-$F_4$ minimum is comparatively shallow for both $G_{110}$ and $G_{100}$. Unlike all other rate minima involving alkane crystallization from octacosane, no needle-like crystals appear at the $F_3$-$F_4$ transition, since neither $G_{110}$ nor $G_{100}$ are hindered drastically. The explanation of the shallow minimum can be inferred by considering the diagram in the inset on the right of Fig. 34. While the difference in stem length between the E and $F_2$ forms is $l^E - l^{F_2} = 20.8$ nm, and while $l^{F_2} - l^{F_3} = 7.0$ nm, the difference $l^{F_3} - l^{F_4}$ is only 3.5 nm. It was argued above that the key cause of self-poisoning is the fact that depositing a longer stem onto a shorter one is unproductive, and that such deposition must be reversed before productive deposition on the growth face can proceed. However, there must be a lower limiting value of the difference $l^{F_x} - l^{F_y}$ below which this condition is not strictly obeyed, and it appears that 3.5 nm is below this value. In other words, a certain proportion of stems that fall somewhat short of the required length are incorporated and tolerated behind the growth front. This implies a degree of roughness of the basal surface. Such roughness has indeed been verified independently, as described below.

## 5.3
### Perpendicular vs. Tilted Chains and Surface Roughness

In crystalline polymers, chains are often tilted relative to the layer normal. The development of tilt is associated with crystallization or annealing at elevated temperatures. Thus in some cases single crystals of a polymer with an orthogonal or nearly orthogonal unit cell grown from solution at low temperatures are flat lamellae with chains perpendicular to the basal (001) surface [65, 76]. However, when grown at a higher temperature the crystals are hollow pyramids or chair-like [77]. Direct crystallization from the melt at high $T_c$ normally yields lamellar crystals with tilted chains [78, 79].

Chain tilt helps alleviate the overcrowding problem at the crystal–amorphous interface [80]. In polyethylene and long alkanes crystallized from melt or from poorer solvents at high temperatures, {100} growth sectors prevail and the tilt is usually 35°, since the basal plane is {201}. Such tilt allows chain folds or ends an increased surface area, by a factor of $1/(\cos 35°)$, while maintaining the crystallographic packing of the remaining chain intact.

In apparent agreement with the behaviour of polyethylene and long alkanes, shorter $n$-alkanes crystallize with perpendicular chains provided that $T_c$ is below ca. 60–70 °C [81]. This is true for odd-numbered alkanes, while even-numbered alkanes display a more complex behaviour due to molecular symmetry [82]. At higher temperatures in alkanes such as $C_{33}H_{68}$, a {101} tilted form is brought about through one or several discrete transitions [83]. In this case the tilt at high temperature is associated with equilibrium conformational disorder of chain ends, and the absence of tilt at low temperature with high end-group order.

Crystallization of extended or folded long alkanes from good solvents at relatively low $T_c$ yields perpendicular chains (see, e.g., Fig. 2). At elevated temperatures chain tilt appears in dried crystals; the tilt angle increases with temperature and usually reaches 35° below the melting point (see, e.g., Fig. 35) [84].

The alkane used in the SAXS experiment in Fig. 35 is extended-chain solution-crystallized $C_{216}H_{385}D_{49}$, having a deuterated $C_{12}D_{25}$ and a $C_{12}D_{24}$ group at the two ends (for synthesis see [85]). This allowed translational disorder at the crystal surface to be studied by IR spectroscopy. If there was perfect order, all $C_{12}D_{25}$ groups would be surrounded by other $C_{12}D_{25}$ groups and CD bending vibration bands would show the full crystal-field splitting of $\Delta\nu = 9.1$ cm$^{-1}$ for perpendicular chains and 8.3 cm$^{-1}$ for $\{201\}$ tilted chains [86]. However, the actual splitting observed in as-grown crystals is less than 7 cm$^{-1}$, with a singlet appearing between the two components (see Fig. 36). This indicates significant random shear between neighbouring chains, or translational disorder, as illustrated schematically in Fig. 37a. Quantitative estimatation gives the size of the average ordered domain as ca. $3 \times 3 = 9$ chains in the as-crystallized sample (from $\Delta\nu$), with about one in nine chains being displaced longitudinally by a distance of the order of 12 C atoms (from the intensity of the singlet) [86].

After annealing at increasing temperatures, simultaneously with chains tilting, the splitting of the CD band increases and the singlet component diminishes, indicating an improvement in translational chain-end order. After annealing at 125 °C the size of the ordered domains increased to $6 \times 6 = 36$ chains and there were almost no large excursions of chain ends left. The smoothening of the basal surface and the introduction and increase of chain

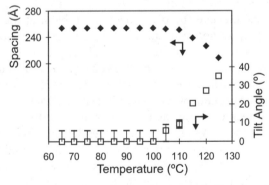

**Fig. 35** Temperature dependence of lamellar spacing (*upper half*) and angle of chain tilt with respect to the lamellar normal (*lower half*) for extended-chain crystals of end-deuterated alkane $C_{216}H_{385}D_{49}$ grown from toluene solution at 70 °C. SAXS spectra were recorded during heating from 60 °C to the melting point. Data for heating at 1 °C/min (*solid diamonds*) and 6 °C/min (*open squares*) are shown for comparison (from [84] by permission of American Chemical Society)

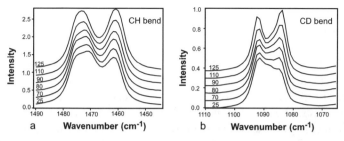

**Fig. 36** IR spectra in (**a**) the $CH_2$ bending region and (**b**) $CD_2$ bending region for extended-chain $C_{216}H_{385}D_{49}$ crystals as-grown from solution and annealed for 30 min at the temperatures indicated (in °C). The spectrum of unannealed crystals is labelled "25". All spectra were recorded at − 173 °C (from [84] by permission of American Chemical Society)

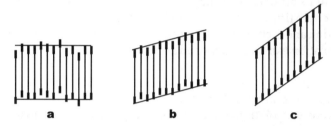

**Fig. 37** Schematic drawing of molecular arrangements in extended-chain crystals of $C_{216}H_{385}D_{49}$ (**a**) as grown from solution (**b**) annealed at a low temperature–low tilt (**c**) annealed at a high temperature–high tilt. *Thickened lines* indicate deuterated chain ends (from [84] by permission of American Chemical Society)

tilt (Fig. 37b,c) are thought to be linked. Thus, in the initial rough surface the overcrowding problem is not serious since the dissipation of crystalline order is gradual. As the density gradient becomes steeper on annealing, overcrowding poses a problem which is alleviated by chain tilting.

Quenching of alkanes such as $C_{162}H_{326}$ from the melt at sufficiently large supercooling also produces crystals with initially perpendicular chains, with tilting occurring soon after [86]. This is in agreement with the morphological study on melt-crystallized polyethylene [87] where it was proposed that below $T_c \cong 127\,°C$ chains may be initially perpendicular, with fold ordering and chain tilting occurring subsequently. This has been offered as an explanation for the S profile of the lamellae as viewed along the growth (*b*) axis.

The kinetic roughness of the basal surface in as-grown crystals ties well with the interpretation of the shallow rate minimum at $T_c^{F_3-F_4}$, described in the preceding section. It is reasonable that some of the 3.5-nm shorter F4 stems would be incorporated into F3 crystals, especially since half of this shortfall in length, i.e. 1.75 nm, could be allocated to each crystal surface. This value is of the same order of magnitude as some of the large excursions of chains detected in as-grown crystals by IR spectroscopy.

# 6
# Conclusions

While the self-poisoning effect leading to the anomalous minimum in temperature dependence of crystallization rate of long alkanes had already been discovered in 1987, extensive recent work has greatly extended our understanding of this effect. The phenomenon itself is only an extreme manifestation of the pinning effect in polymer crystallization. It is only due to the quantized nature of lamellar thickness ("integer folding") in monodisperse alkanes that the effect has become visible, although there is now strong experimental indication that self-poisoning also plays an important role in crystallization of polyethylene, particularly at higher temperatures, and that it is largely responsible for the shape of its crystals (Sect. 4.2). Recent new information on self-poisoning has emerged particularly thanks to direct in situ measurements of growth rates from solution and melt combined with the analysis of crystal habits. Briefly, the new information can be summarized as follows:

1.  A minimum in growth rate occurs as a function of either crystallization temperature or solution concentration at all growth transitions between successive integer folded forms. The latter results in negative-order kinetics, the dilution wave effect, and autocatalytic crystallization.
2.  Both rate dependencies can be explained theoretically using simple analytical models as well as computer simulation. The only necessary prerequisites are the consideration of competing chain depositions and the inability of a longer stem to attach to an underlying shorter stem.
3.  Additional rate minima can be created in melt crystallization within the extended-chain growth regime, by adding a selected shorter-chain guest alkane ("poisoning minima"). This supports the interpretation of the rate anomalies in long alkanes in terms of obstructing short stem attachments.
4.  Retardation of crystal growth due to self-poisoning occurs preferentially on $\{100\}$ faces.
5.  Step initiation (secondary nucleation) on $\{100\}$ faces virtually stops at the rate minimum, while a smaller decrease occurs in step propagation rate $v$. The observed minimum in $v$ is inconsistent with the barrier to stem deposition being determined by surface energy and suggests that, as in secondary nucleation, the barrier is entropic.
6.  Self-poisoning is more prominent where stem length is longer.
7.  Self-poisoning is more prominent in solvents consisting of smaller nonlinear aromatic molecules as compared to the linear aliphatic solvent octacosane. The effect may be due to varying degrees of molecular preordering in solution.
8.  There is experimental evidence in favour of two-dimensional secondary nucleation of chain-folded crystals.

9. The results generally support the secondary nucleation nature of polymer crystallization, but highlight the importance of considering competing attachments and pinning. Step propagation can be highly retarded due to self-poisoning so that in extreme cases nucleation control becomes negligible, with the limit of rough-surface growth being reached.
10. The rate minimum is shallower if the obstructing stems are close in length to those of the substrate. This is consistent with the tolerance of limited fluctuation in chain end location by kinetically rough basal lamellar surfaces.

Finally, it should be mentioned that self-poisoning due to "erroneous" integer folded depositions is only one manifestation of a productive reaction, which leads to thermodynamic stability, being retarded by a competing lower barrier reaction which *almost* leads to a stable product. The observations of a crystallization rate minimum in an aromatic polyketone [88], and recently an aromatic polyester [89], are further examples of such an occurrence.

**Acknowledgements** The authors are grateful for financial support from the Engineering and Physical Science Research Council and from INTAS.

## References

1. Brooke GM, Burnett S, Mohammed S, Proctor D, Whiting MC (1996) J Chem Soc Perkin Trans 1:1635
2. Lee KS, Wegner G (1985) Makromol Chem Rapid Commun 6:203
3. Brooke GM, Burnett S, Mohammed S, Proctor D, Whiting MC (1996) J Chem Soc Perkin Trans 1:1635
4. Brooke GM, Mohammed S, Whiting MC (1999) Polymer 49:773
5. Atkins EDT, Hill MJ, Jones NA, Sikorski P (2000) J Mater Sci 35:5179
6. Organ SJ, private communication
7. Ungar G, Zeng XB (2001) Chem Rev 101:4157
8. For a review of polymer crystallization theory see: Armitstead K, Golbeck-Wood G (1992) Adv Polym Sci 100:219
9. Ungar G, Stejny J, Keller A, Bidd I, Whiting MC (1985) Science 229:386
10. Magonov SN, Yerina NA, Ungar G, Reneker DH, Ivanov DA (2003) Macromolecules 36:5637
11. Ungar G, Keller A (1986) Polymer 27:1835
12. Ungar G, Zeng XB, Brooke GM, Mohammed S (1998) Macromolecules 31:1875
13. Zeng XB, Ungar G, Spells SJ (2000) Polymer 41:8775
14. Zeng XB, Ungar G, Spells SJ, Brooke GM, Farren C, Harden A (2003) Phys Rev Lett 90:155508
15. Zeng XB, Ungar G, Spells SJ, King S (2004) Real-time SANS study of transient phases in polymer crystallization. Highlights of ISIS Science, Ann Rep Rutherford-Appleton Lab (in press)
16. Hoffman JD, Davis GT, Lauritzen JI Jr (1976) In: Hannay NB (ed) Treatise on solid-state chemistry, vol 3. Plenum, New York, pp 497–614
17. Hoffman JD (1997) Polymer 38:3151
18. Kovacs AJ, Gonthier A, Straupe C (1975) J Polym Sci Polym Symp 50:283

19. Hoffman JD (1986) Macromolecules 19:1124
20. Hoffman JD (1985) Macromolecules 18:772
21. Sadler DM (1985) J Polym Sci Polym Phys 23:1533
22. Sadler DM (1983) Polymer 24:1401
23. Point JJ, Kovacs A (1980) Macromolecules 13:399
24. Toda A (1986) J Phys Soc Jpn 55:3419
25. Ungar G (1993) In: Dosiere M (ed) Polymer crystallization. NATO ASI Series. Kluwer, Dordrecht, pp 63–72
26. Ungar G, Keller A (1987) Polymer 28:1899
27. Organ SJ, Ungar G, Keller A (1989) Macromolecules 22:1995
28. Organ SJ, Keller A, Hikosaka M, Ungar G (1996) Polymer 37:2517
29. Putra EGR, Ungar G (2003) Macromolecules 36:5214
30. de Silva DSM, Ungar G, in preparation
31. de Silva DSM, Zeng XB, Ungar G, in preparation
32. Organ SJ, Keller A (1985) J Mater Sci 20:1571
33. Toda A (1991) Polymer 32:771
34. Boda E, Ungar G, Brooke GM, Burnett S, Mohammed S, Proctor D, Whiting MC (1997) Macromolecules 30:4674
35. Organ SJ, Barham PJ, Hill MJ, Keller A, Morgan RL (1997) J Polym Sci Polym Phys 35:775
36. Sutton SJ, Vaughan AS, Bassett DC (1996) Polymer 37:5735
37. Hobbs JK, Hill MJ, Barham PJ (2001) Polymer 42:2167
38. Morgan RL, Barham PJ, Hill MJ, Keller A, Organ SJ (1998) J Macromol Sci Phys B 37:319
39. Putra EGR, Ungar G (2003) Macromolecules 36:3812
40. Cheng SZD, Chen JH (1991) J Polym Sci Polym Phys 29:311
41. Hosier IL, Bassett DC, Vaughan AS (2000) Macromolecules 33:8781
42. Zeng XB, Ungar G (2001) Macromolecules 34:6945
43. Zeng XB, Ungar G, Spells SJ, Brooke GM, Farren C, Harden A (2003) Phys Rev Lett 90:155508
44. Ungar G, Mandal P, Higgs PG, de Silva DSM, Boda E, Chen CM (2000) Phys Rev Lett 85:4397
45. Putra EGR, Ungar G, in preparation
46. Higgs PG, Ungar G (2001) J Chem Phys 114:6958
47. Fisher RA (1937) Ann Eugen 7:355
48. Sadler DM, Gilmer GH (1987) Polym Commun 28:243
49. Higgs PG, Ungar G (1994) J Chem Phys 100:640
50. Keith HD (1964) J Appl Phys 35:3115
51. Bassett DC, Hodge AM (1981) Proc R Soc Lond A 377:25
52. Ungar G, Putra EGR (2001) Macromolecules 34:5180
53. Waddon AJ, Ungar G, in preparation
54. Khoury F (1979) Faraday Discuss Chem Soc 68:404
55. Mansfield ML (1988) Polymer 29:1755
56. Point JJ, Villers D (1991) J Cryst Growth 114:228
57. Frank FC (1974) J Cryst Growth 22:233
58. Hoffman JD, Miller RL (1989) Macromolecules 22:3038, 3502; Hoffman JD, Miller RL (1991) Polymer 32:963
59. Toda A (1993) Faraday Discuss 95:129
60. Shcherbina M, Ungar G, in preparation
61. Toda A (2003) J Chem Phys 118:8446
62. Sadler DM, Gilmer GH (1988) Phys Rev B 38:5684

63. Buckley CP, Kovacs AJ (1984) In: Hall IH (ed) Structure of crystalline polymers. Elsevier, London, pp 261–307
64. Kovacs AJ, Gonthier A (1972) Kolloid Z Z Polym 250:530
65. Briber RM, Khoury F (1993) J Polym Sci Polym Phys 31:1253
66. Toda A, Arita T, Hikosaka M (2001) Polymer 42:2223
67. Pechhold W (1967) Kolloid Z Z Polym 216:235
68. Heck B, Strobl G, Grasruck M (2003) Eur Phys J E 11:117
69. Putra EGR, Ungar G, submitted
70. Point JJ (1979) Faraday Discuss Chem Soc 68:167
71. Doye JPK, Frenkel D (1998) J Chem Phys 109:10033
72. Hikosaka M (1987) Polymer 28:1257; Polymer (1990) 31:458
73. Welch P, Muthukumar M (2001) Phys Rev Lett 87:218302
74. Frank FC, Tosi M (1961) Proc R Soc Lond A 263:323
75. Sadler DM, Gilmer GH (1984) Polymer 24:1446
76. Okuda K, Yoshida T, Sugita M, Asahina M (1967) J Polym Sci A-2 5:465
77. Bassett DC, Frank FC, Keller A (1963) Phil Mag 8:1753
78. Khoury F (1979) Faraday Discuss Chem Soc 68:404
79. Bassett DC, Hodge AM (1981) Proc R Soc Lond A 377:25
80. Guttman CM, DiMarzio EA, Hoffman JD (1981) Polymer 22:1466
81. Smith AE (1953) J Chem Phys 21:2229
82. Shearer HMM, Vand V (1956) Acta Crystallogr 9:379
83. Piesczek W, Strobl GR, Malzahn K (1974) Acta Crystallogr B 30:1278
84. de Silva DSM, Zeng XB, Ungar G, Spells SJ (2002) Macromolecules 35:7730
85. Brooke GM, Farren C, Harden A, Whiting MC (2001) Polymer 42:2777
86. de Silva DSM, Zeng XB, Ungar G, Spells SJ (2003) J Macromol Sci Phys 42:915
87. Abo el Maaty MI, Bassett DC (2001) Polymer 42:4957
88. Blundell DJ, Liggat JJ, Flory A (1992) Polymer 33:2475
89. Ten Hove CF, De Meersman B, Penelle J, Jonas AM (2004) Proc IUPAC Macro 2004, Paris

Adv Polym Sci (2005) 180: 89–159
DOI 10.1007/b107233
© Springer-Verlag Berlin Heidelberg 2005
Published online: 29 June 2005

# Effect of Molecular Weight and Melt Time and Temperature on the Morphology of Poly(tetrafluorethylene)

P. H. Geil[1] (✉) · J. Yang[1,2] · R. A. Williams[1] · K. L. Petersen[1] · T.-C. Long[3] · P. Xu[3]

[1]Department Of Materials Science And Engineering, University Of Illinois, Urbana, IL 61801, USA
*geil@uiuc.edu, junyan@engin.umich.edu, rawilli1@uiuc.edu, klpeter1@uiuc.edu*

[2]Department Of Materials Science And Engineering, University Of Michigan, Ann Arbor, MI 48109, USA
*junyan@engin.umich.edu*

[3]W. L. Gore and Assocs., Ltd., Newark, DE  USA
*tlong@wlgore.com, pxu@wlgore.com*

**Abstract** Evolution of the morphology of dispersed particles of various molecular weight poly(tetrafluorethylene dispersion particles has been examined as a function of sintering time and temperature. Substantial molecular motion on the substrate occurs: large, angular particles form first for short melt times, followed by development of both planar, folded-chain single crystals, and single-molecule single crystals and banded structures with parallel double striations oriented along their long axis. The molecules in the latter two are parallel to the substrate. Deformation of the particles results in indefinite length nanofibrils; sintering results in the growth of shish kebabs, again by individual molecule mobility. Tentatively it is suggested that the morphology develops in a mesomorphic "melt", the morphology depending primarily on sintering time and temperature rather than the rate of cooling. Consideration is also given to the morphology of the nascent particles; chain folding during polymerization is indicated for nanoemulsion particles.

**Keywords** Bands · Morphology · Nanofibrils · Poly(tetrafluorethylene) · Single crystals · Shish kebabs

**Abbreviations**

AFM   Atomic force microscopy
BFDC  Bright-field diffraction contrast
DFDC  Dark-field diffraction contrast
ED    Electron diffraction
PTFE  Poly(tetrafluoroethylene)
SEM   Scanning electron microscopy
TEM   Transmission electron microscopy

# 1
# Introduction

The banded structure on fracture surfaces of poly(tetrafluoroethylene) (PTFE) crystallized slowly from the melt has been known for more than 40 years [1–4]. The PTFE bands were first observed [1] by melting compacted granular and dispersion[1] PTFE at 380 °C (melt time not given), followed by slow cooling. The bands seen on fracture surfaces of the granular material, having striations on their surfaces normal to the bands (Fig. 1), closely resemble, in appearance and probably growth mechanism, the extended-chain lamellae seen on fracture surfaces of linear polyethylene crystallized slowly at elevated pressure (see Fig. 1 in Ref. [5]); in lower molecular weight material they appeared to be arranged radially in spherulitic structures [1, 4, 6]. However, because of difficulties in characterizing the molecular weight of PTFE, it was not known if the bands were extended-chain (thickness equals chain length, with its implications for molecular weight fractionation during crystallization) or chain-extended (thickness less than the chain length, and therefore folded, but greater than several hundred angstroms)[2]. Bunn et al. suggested the former, with fractionation by molecular weight carried over

---

[1] Also described as emulsion PTFE; it is produced by an emulsion polymerization process. The granular material is polymerized in the bulk, the nascent particles being much larger. (see Figs. 8, 9 in Ref. [4])

[2] Definitions we use. Extended chain is also defined as fully extended. But also see Ref. [7], which suggests these terms have been used in various contexts in polymer science, resulting in a certain amount of confusion

**Fig. 1** Fracture surface of granular (bulk-polymerized) poly(tetrafluorethylene (*PTFE*). The sample was held in the melt at 380 °C for an unspecified time followed by slow cooling and fracture after immersion in liquid $N_2$. (Reprinted from Ref. [1] with permission from Wiley-Interscience)

from the polymerization, even though the band thickness corresponded to a chain length with a molecular mass about 10% of the estimated $3 \times 10^6$ Da from the incorporation of radioactive ends. They suggested "in the melt" disorder is very far from complete, and that the molecules are straighter and less tangled than in most polymer melts, but also are in rough bundles with their ends by no means randomly placed. On the other hand, rapid cooling of the melt resulted in irregular fracture surfaces and, from X-ray measurements, smaller crystals and lower crystallinity. They thus suggested "the band structures are due to the processes of crystal growth, involving a certain amount of molecular movement, though large migrations are unlikely." The striations were attributed to either slip planes or repeated twinning.

The dispersion material, of lower molecular weight, also showed similar thickness bands on fracture surfaces, except that, in this case, the striations were parallel to the length of the bands and very regularly spaced at about 1000 Å (Fig. 2). The central band, extending from lower left to upper center has the appearance of having split during fracture, with the "subsheets extending from one level to another" (added arrows). Although submitted about 1 year after Keller's paper proposing chain folding in solution-grown polyethylene crystals [8], the possibility of chain folding to explain the 1000-Å thickness was considered more improbable than a uniform molecular weight. We see similar structures on free surfaces of thin-film, melt-crystallized dispersion material (Figs. 24b, c, 25a, 26d, 27b, 28a) but have never observed them on fracture surfaces; thus, we suggest this micrograph

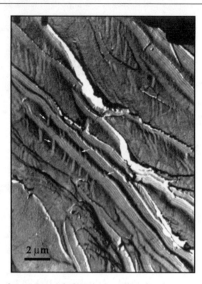

**Fig. 2** Fracture surface of compacted dispersion PTFE. The treatment was similar to that in Fig. 1. The *arrow* indicates a region where a striation terminates at the edges of a band, suggesting the striations are edges of lamellae. (Reprinted from Ref. [1] with permission from Wiley-Interscience)

of Bunn et al. may have resulted from surfaces of interior voids. Using slow-cooled, radiation-degraded PTFE they observed thin lamellae, approximately 200-Å thick, on free surfaces (Fig. 3). Again suggesting they were chain-extended crystals, they acknowledged the difficulty of explaining the then required uniformity of chain length in a degraded material.

The report by Bunn et al. [1] was followed, in 1962, by two papers by Speerschneider and Li [2, 3] describing bands on fracture surfaces of compacted dispersion PTFE similar to those shown in Fig. 1 for granular material. They also used a melt temperature of 380 °C, with the melt time not given. These authors described the effect of deformation at various temperatures. They suggested the striations had a regular lateral spacing of 200 Å regardless of the rate of cooling the sample, whereas Bunn et al. indicated the spacing varied widely, down to approximately 300 Å (both sets of authors used a two-stage replication technique, limiting the resolution). Speerschneider and Li suggested the striations were the edges of sheets corresponding to alternating crystalline and amorphous layers, totaling 200 Å, with deformation giving rise to slipping of the layers in the amorphous regions at elevated temperatures.

At about the same time as the papers by Speerschneider and Li, Symons [4, 9] described the growth of lamellar single crystals from dispersed PTFE dispersion particles (Fig. 4) and we described thin lamellae on the free surfaces of bulk dispersion PTFE samples (Fig. 5) [4], both prepared by slow cooling fol-

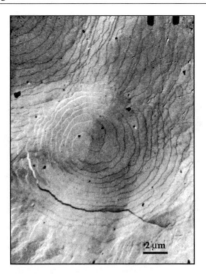

**Fig. 3** Free surface of radiation-degraded PTFE after being held in the melt at 380 °C and slow cooling. The lamellae are approximately 200-Å thick based on the shadowing. (Reprinted from Ref. [1] with permission from Wiley-Interscience)

lowing 2-h melt times at 350 °C (Fig. 4) or from 380 °C (Fig. 5). For both types of samples chain folding was proposed, the melt temperature being believed to be too low for extensive degradation. The surface of the lamellae in Fig. 5 is irregular. One possibility is that they are cracks spanned by fibrils, with the cracks and fibrils resulting from thermal shrinkage and unfolding of the chains. Another possibility is that the "cracklike" structures resulted from

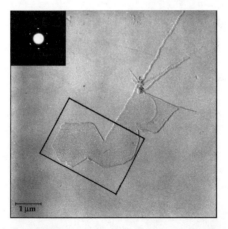

**Fig. 4** Lamellar, folded-chain single crystals of dispersion PTFE prepared by sintering dispersed dispersion particles at 350 °C for 2 h on a glass slide. (Reprinted from Ref. [9] with permission from Wiley-Interscience)

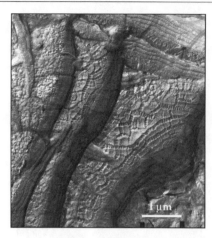

**Fig. 5** Lamellae on a free surface of compacted PTFE dispersion particles sintered at 380 °C followed by slow cooling. (Similar to Fig. IV-67in Ref. [4])

annealing of fibrils drawn out and deposited on the surface during fracture, resulting in shish kebabs. They resemble the structures described by Wunderlich and Melilio [10] for annealed fibrils on the 001 (chain-end) surfaces on fracture surfaces of pressure-crystallized, extended-chain linear polyethylene, the fibrils having been formed during fracture.

Individual bands were also seen on the free surfaces of both granular and dispersion bulk samples; irregular continuous striations were present parallel, rather than normal, to the long axes of the bands (Fig. 6). Bands with well-

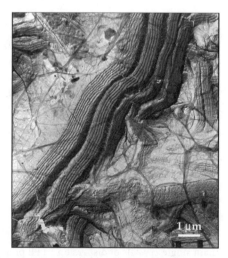

**Fig. 6** Bands on the free surface of compacted PTFE dispersion particles sintered at 380 °C followed by slow cooling. (Similar to Fig. IV-65 in Ref. [4])

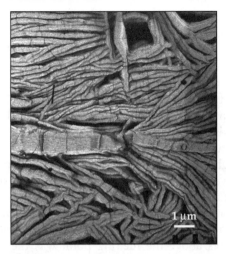

**Fig. 7** Bands on the free surface of compacted low molecular weight PTFE dispersion particles. (Similar to Fig. IV-66 in Ref. [4])

formed striations parallel to the long axis, similar to those described by Bunn et al. for dispersion PTFE (Fig. 2) [1], were observed on the free surface of a bulk, low molecular weight dispersion PTFE (Fig. 7) [4], upper right of Fig. 6, and by Symons [9] in thicker regions of the dispersed dispersion particles (Fig. 8). In Figs. 4 and 7 the striations consist of double lines, a characteristic discussed further later.

Fracture surfaces of both dispersion and granular high molecular weight PTFE that we described [4] had bands similar to those described by Bunn

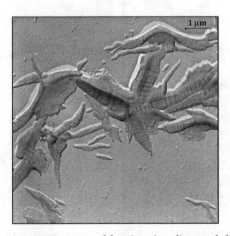

**Fig. 8** Bands of dispersion PTFE prepared by sintering dispersed dispersion particles at 350 °C for 2 h on a glass slide. (Reprinted from Ref. [9] with permission from Wiley-Interscience)

et al. [1], i.e., with striations normal to the bands. In a possibly related paper, Wunderlich and Melilio [6] showed that striations oriented parallel to the long axes of the bands, but shorter and more irregular, could be grown on the surfaces of Bunn et al. type bands on fracture surfaces of slow-cooled PTFE by subsequent annealing at 315–326 °C for 30 s followed by ice–water quenching (also better formed in a sample held at 316 °C for 40 min, Fig. 9); they suggested their striations consisted of folded-chain lamellae developing by epitaxial crystallization on the extended-chain bands. Possibly the same effect could have occurred on the free surfaces during crystallization, exuded low molecular weight material crystallizing epitaxially during cooling on the higher molecular weight bands. Thus, correlation of the structure on PTFE-free and fracture surfaces remains unclear.

Another observation by Wunderlich and Melilio [6] of both interest and concern relative to our observations later of individual molecules moving on the substrates above the melting point was the development of fibrous, presumably extended-chain structures by monomer sublimation and repolymerization/crystallization on the walls of an evacuated tube when a PTFE sample in it was heated to 320 °C for 2 h (Fig. 10). Butenuth [11] had earlier described similar observations. On colder surfaces the polymer formed as strings of less than 0.1-micron circular platelets that the authors suggested formed as a shish-kebab structure (Fig. 10, right side).

The morphology of nascent PTFE dispersion also remains unclear. Several authors, during the next decade of PTFE morphology research, suggested standard-size PTFE dispersion particles (ellipsoidal particles with dimensions

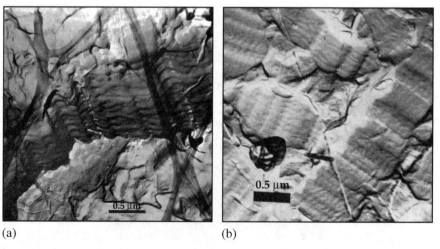

(a)                                                                                    (b)

**Fig. 9** Fracture surfaces of slow-cooled (at 4.6 °C/h after 5 h at 355 °C) PTFE samples that were subsequently annealed at 316.2 °C for 40 min and 319.2 °C for 30 s. (Reprinted from Ref. [6] with permission from Springer-Verlag)

**Fig. 10** PTFE filaments prepared by heating a PTFE sample in an evacuated tube to 320 °C for 2 h. The filaments collected on the hot wall of the tube. The *inset* shows the "shish-kebab" structure formed when the walls were cool. (Reprinted from Ref. [6] with permission from Springer-Verlag)

greater than 1000 Å) consist of folded ribbons [1, 12–14]. On the other hand, we proposed that the folded-ribbon appearance was due to beam damage (Fig. 15, [15, 16]) and, on the basis of dark-field diffraction contrast (DFDC) and bright-field diffraction contrast (BFDC) micrographs of Teflon[3] dispersion particles (Fig. 14), suggested [4] that the dispersion particles contained "pie-shaped" crystals, the molecules being tangential in the particles. As described by Seguchi et al. [14], PTFE dispersion particles (described as granules), depending on the polymerization conditions, can have a shape varying from fibrils to rods to elliptical particles (Fig. 11). The fibrils were as small in diameter as 200 Å and microns in length, the rods were 300–600 Å in diameter and shorter than the fibrils, while the elliptical particles were still larger in diameter and shorter, resembling the commercial material. They suggested the variation in shape corresponded to a variation in molecular mass, from below $2 \times 10^4$ Da for the filaments, from $2 \times 10^5$ to $5 \times 10^5$ Da for rods, to above $10^6$ Da for the elliptical particles, with the variation in shape and molecular weight related to changes in emulsifier concentration. The filaments formed at a 2% concentration, the rods at 1%, a mixture of rods and particles at 0.5% and at 0% only nearly spherical to elliptical particles. They suggested the fibrils folded to form the elliptical particles, in a manner similar to that proposed earlier by Rahl et al. [13] and Bunn et al. [1], both of whom proposed the elliptical particles were formed by a folding up of ribbons. Seguchi et al. [14] suggested the filaments were extended-chain structures with the 500-Å long chains being considerably shorter than the greater

---

[3] Trademark of E.I. du Pont de Nemours for its PTFE resin

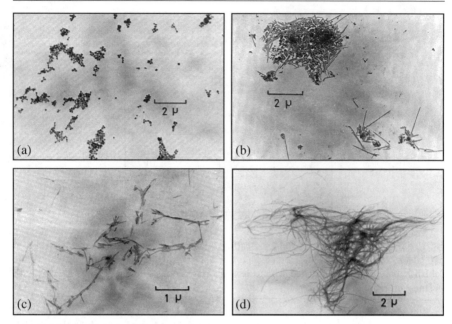

**Fig. 11** Nascent radiation initiated PTFE dispersion particles. The polymerization conditions were (**a**) 0% emulsifier, 90 min, (**b**) 0.5% emulsifier, 60 min, (**c**) 1% emulsifier, 60 min, (**d**) 2% emulsifier, 90 min, all for essentially the same radiation dose rate at 25 °C in water (with 1.3% hexadecane and ammonium perfluorooctanoate emulsifier) at 30-kg/cm² pressure. The measured molecular weights, the corresponding extended chain lengths (*ecl*) the and dimensional characteristics are, respectively, (**a**) $\overline{M}_n = 230 \times 10^4$, $ecl_n = 6.0$ μm, particle volume approximately $7 \times 10^8$ Da; (**b**) $\overline{M}_n = 50 \times 10^4$, ecl = 1.3 μm, rod diameter approximately 100 molecules; (**c**) $\overline{M}_n = 20 \times 10^4$, ecl = 0.52 μm ≈ rod length; (**d**) $\overline{M}_n = 2 \times 10^4$, ecl = 520 Å. (Reprinted from Ref. [14] with permission from Wiley-Interscience)

than 2-μm long filaments, resulting in defects and permitting the fibrils to fold to form the rods and elliptical particles. The rod length, however, was about the same as the extended-chain length and their diameters were similar to the diameters of the spherical particles polymerized with 0% surfactant; thus, rod folding to form the particles seems unlikely. Rahl et al. [13], on the other hand, had proposed the elliptical particles were formed from the folding of thin (approximately 60-Å thickness) ribbons of extended-chain molecules, again shorter than the ribbon length. Of particular interest was a micrograph of a mechanically "dismantled" particle in which the ribbon was partially unfolded (Fig. 12); the dismantling was accomplished by drawing a C-coated grid across a drop of the suspension. On the basis of electron diffraction (ED) from the particles and easy cleavage parallel to the axis of the ribbons, they suggested the ribbons consisted of extended chains parallel to the long axis of the ribbons.

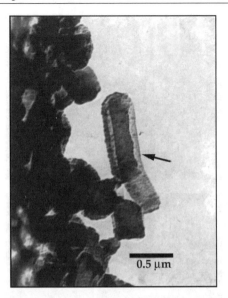

**Fig. 12** "Partially unfolded" particle of dispersion PTFE produced by drawing a C-coated transmission electron microscope grid across the surface of a drop of suspension. (Reprinted from Ref. [13] with permission from Wiley-Interscience)

The structure of the dispersion particles was reexamined in the middle of the 1980s, using high-resolution electron microscopy, by Chanzy and coworkers [17–19]. A diffraction contrast image of elliptical ("spherical") dispersion particles was similar to that in Figs. 1 and 2, pie-shaped sectors corresponding to the crystalline regions (inset in Fig. 13a). A lattice image of a corresponding sector is shown in Fig. 13a [17], with the optical diffraction pattern shown in the top right inset. Both the ED pattern corresponding to the diffraction contrast inset and the optical diffraction pattern indicate the molecular axes are tangential in the imaged sectors. Lattice images of rod [17] (Fig. 13b) and fibril (also called rods) particles [19] also revealed the presence of coherent crystalline regions, in this case $10^3$–$10^4$ Å in length, with the chain axis parallel to the rod axis. They also suggested [17, 19] that the rods (or fibrils) folded back on themselves to form the elliptical particles, considering the rods to be whiskers (i.e., single crystals). Similar to Seguchi et al. [14], the folding was attributed to interfacial forces developing when the surfactant was used up, high surfactant concentration protecting and preserving the whiskers.

Using oligomeric dispersion particles (estimated molecular weight of 3000 based on the melting point, i.e., 60 C atoms corresponding to an extended chain length of approximately 80 Å) that were more or less hexagonal platelets of varying thickness extending up from ca. 50 Å[4], Chanzy et al. [18]

---

[4] The paper lists 50 nm, but says this corresponds to 40 C atoms.

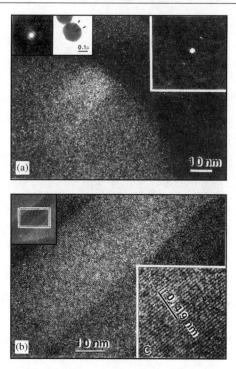

**Fig. 13** Lattice images of (**a**) a sector in an elliptical particle and (**b**) a section of a rod particle. The *upper-left inset* in **a** is the electron diffraction (*ED*) pattern of the selected area of the diffraction contrast image; two sectors contribute to the image with the reflections being 100. The *upper-right inset* is the optical diffraction pattern from the bright area in the image. In **b** the *upper-left inset* is a low-magnification image of the section of the rod selected, while the *lower-right inset* is a magnified portion of the lattice image; the spacings correspond to 100. (Modified from Ref. [17] with permission from Wiley-Interscience)

were able to image the hexagonal lattice, enhanced by computer filtering. An optical diffraction pattern from the original negative had the 100 reflections clearly resolved, with coherent 100 lattice planes (in one direction) extending over distances greater than $10^5$ Å. The relationship of the oligomeric crystals, approximately 1000 Å in diameter, and the whiskers, approximately 200 Å in diameter, to each other and to the growth of the elliptical particles was not described. The surfactant concentration used for growth of the oligomeric crystals was not stated but presumably was low since they were extracted from a "conventional" PTFE polymerization at an early stage (less than 1% conversion) [17], with conventional dispersions being predominantly elliptical particles by design.

Considered here is the morphology of two commercial, so-called nanoemulsion PTFEs (specially prepared such that at least one dimension is below 1000 Å, both rodlike and hexagonal in shape), and the morphologies obtained by melting and then crystallizing their dispersed particles on a substrate as a function of melt time and temperature. For comparison, similar studies with several standard-size PTFE dispersions are also described. Although we discuss the structure of the particles and the morphology of the material crystallized from the bulk melt, they remain items of concern and interest.

# 2
# Experimental

The majority of the research described here was done with nanoemulsions obtained from DuPont (TE-5070) and Ausimont (18749/26, hereafter A18749). $\overline{M}_n$ of TE-5070, determined from the zero-shear melt viscosity, was 50 000 Da, corresponding to an extended-chain length of 1300 Å, with a degree of polymerization of 540.[5] The light-scattering-determined particle size was 650 Å. The A18749 nanoemulsion had $\overline{M}_n = 200 000$ Da, $\overline{M}_w = 49 \times 10^6$ Da (number-average chain length of 5200 Å, weight-average chain length of $1.3 \times 10^6$ Å $= \underline{13 \text{ mm}}$) and a particle size of 335 Å. The standard-size dispersion particles were several products of DuPont, TE-30, TE-3170 and TE-3698, and a special resin prepared for W.L. Gore (labeled sample G) in the form of dispersions, with sample G also available in powder form. The measured weight-average molecular mass of TE-30 was $15-20 \times 10^6$ Da, the chain length being $0.4-0.5 \times 10^6$ Å, and sample G had a weight-average molecular mass of $\overline{M}_w$ of $50 \times 10^6$ Da and a weight-average chain length of $1.3 \times 10^6$ Å. The molecular masses of TE-3170 and TE-3698 are assumed also to be above $10^6$ Da.

Diluted virgin dispersion particle suspensions were dried on carbon- or Formvar-coated glass slides, by ambient- or freeze-drying, and shadowed with Pt/C or Cr if desired; the substrates were then floated off on water or dilute HF and picked up on 200 or 400 mesh Cu electron microscope grids. Micrographs of unshadowed samples, in both bright and dark field, were taken under low-beam conditions with a Philips CM 12. Scanning electron microscopy (SEM) photographs of Cr-coated particles of as-cast resin G, and Teflon 30 after a compressive short shear on the glass slide, were taken at 2 kV on a field-emission Hitachi 4700. The as-cast particles on glass slides were also sheared long distances with another slide or razor blade for transmission electron microscopy (TEM). All original TEM micrographs, unless noted, are printed here as negatives, prints being prepared from scans of the negatives. Melting and recrystallization of dispersed dispersion particles was accomplished by drying diluted suspensions on glass cover slips or thin films of mica and heating between the platens of a compression-molding press;

---

[5] In the hexagonal form, phase IV, that is stable from approximately 19 to 35 °C, depending to some extent on molecular weight, there are 38.5 Da/Å chain length and 1.3 Da/Å$^3$ of the lattice; dimensions from Ref. [20]

temperature control was no better than $\pm 5\,°\text{C}$. It is not believed that the individual molecule motion on substrates described later is due to the sublimation repolymerization/crystallization process described by Butenuth [11] and Wunderlich and Melilio [6] since overlying cover slips showed no evidence of transferred material. Furthermore Siegle et al. [21] list rate constants of approximately $8.6 \times 10^{-9}$ at $380\,°\text{C}$, the maximum temperature we used, and less than $2.8 \times 10^{-6}$ at $450\,°\text{C}$ during vacuum pyrolysis of thin PTFE films and Shulman [22] suggests degradation starts at about $450\,°\text{C}$ on the basis of mass spectrometric thermal analysis. Although the degradation rate results suggest little degradation under our sintering conditions, they involve measurements of the production, primarily, of $C_2F_4$ monomer, rather than changes in molecular weight, and would seem incompatible with the TEM results in Refs. [6, 11]. We thus suggest further examination is warranted, particularly since our annealing temperatures and, sometimes, times were greater than those used in the latter references. Sheared samples were "annealed" similarly. Resulting samples for TEM were (usually, unless noted) shadowed with Pt/C, coated with C, floated off on dilute HF (which dissolves the glass or mica) and picked up on grids. Thick samples were prepared from dried dispersions by compressing in a Fourier transform IR KBr pellitizer, followed by heating, in the press, in a mold with holes of the same diameter. Fracture was in liquid nitrogen. The mold was thicker than the pellet, resulting in a free surface.

X-ray diffraction scans at various temperatures were taken at Oak Ridge National Laboratories using an mBraun linear position sensitive detector mounted on a Scintag PAD X $\theta$–$\theta$ diffractometer with a 2-kW Cu $K_\alpha$ source operating at 45 kV, 40 mA and a Buhler high-temperature furnace. Temperatures slightly below room temperature were obtained by running cooling water through the coils of the furnace.

# 3
# Results and Discussion

## 3.1
## Nascant Particle Structure

### 3.1.1
### Standard-Size Dispersion Particle Structure

Figure 14a is a typical TEM micrograph of Pt/C shadowed standard-size PTFE dispersion particles, (here sample G) with dark-field images of unshadowed particles in the insets; typical BFDC micrographs of several standard-size DuPont dispersion particles are shown in Fig. 14b and c. The dark-field micrographs, taken with 100 reflections, are similar to the bright-field ones except

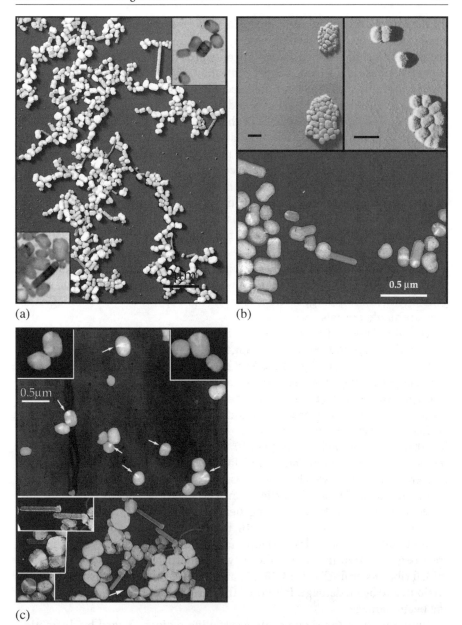

(a)                                        (b)

(c)

**Fig. 14** (**a**) Shadowed sample of dispersed resin G. The *insets* are dark-field images. Low-intensity bright-field diffraction contrast images of (**b**) TE-30 (*bottom*) and (**c**) TE-3698 (*top*), and TE-3170 (*bottom*). The upper part of **b** (**d**) has Pt/C-shadowed TE-30 particles; the *insets* in **c** are enlargements of some of the particles. All *scale bars* in a given figure part represent the same scale unless otherwise identified.

for the reversal of, and increase in, the brightness of the diffracting regions and a reduced number. As shown previously [4, 17], most of the diffracting regions are more or less pie-shaped, extending outwards from the center, if the particle is nearly spherical, or from a short central line, if the particle is elongated. In the micrograph of TE-3698, for instance, pie-shaped domains can be found at the ends of the elongated particles and along their lateral edges (arrows and insets, Fig. 14c,top). The particles in this sample are all spherical to short ellipsoids, with the long axis less than approximately 1.5 times the short axis, while in the other two dispersions the particles vary from spheres to rods. In TE-30 (Fig. 14b) even the nearly equiaxed particles have a short rodlike appearance; the rods, on the other hand, are smaller in diameter (the smallest shown is 630 Å) and appear to grow out from (or into) a nearly spherical particle (arrows). In TE-3170 (Fig. 14c, bottom) most of the particles have a shape similar to that in TE-3698; the few isolated rods, approximately 450 Å in diameter, each have one blunted end, again suggesting linear growth from or folding into a (smaller than for TE-30) nearly spherical particle. Although none of the rods in these bright-field micrographs show diffraction contrast, in other micrographs entire rods or large sections thereof "light up", as in Fig. 13b, suggesting they are single crystals.

The DFDC micrographs (e.g., insets in Fig. 14a) indicate the molecular axes are parallel to the rod axis and are tangential in the pie-shaped domains at the ends of the particles; in Fig. 14a several entire cross-section domains of the rod in the upper right inset are scattering coherently. The length of the longest rod in the micrograph of TE-30 and the longer rods in TE-3170 (inset in Fig. 14c, bottom), assuming they are extended chain crystals, corresponds to molecular masses of approximately 105 000 and 135 000 Da, respectively, considerably less than the presumed greater than $10^6$-Da value. Although they may be a low molecular weight component, chain folding would be required if they consisted of average molecular weight molecules.

All of the undamaged particles in Fig. 14 are of nearly uniform projected electron density, decreasing near the boundaries. There is no TEM evidence of internal structure. With increasing electron beam irradiation, however, all of the particles develop the well-known large differences in contrast (e.g., TE-3170 in Fig. 15) that have been attributed by some authors to folded ribbons or fibrils [1, 13, 14, 17] and that we [4] and others [15, 16] have attributed to beam damage. It is noted the particles shrink considerably during the beam damage.

On the basis of the shape of the crystalline regions imaged by diffraction contrast, if folded ribbons or rods are present in the ellipsoidal particles there must be only a single fold per particle diameter (i.e., three total folds, four rod segments), the ribbon or the rod being half the diameter of the particle. In addition the crystalline regions in neighboring ribbons in some of the particles must be in diffraction register, the crystalline domain spanning the particle,

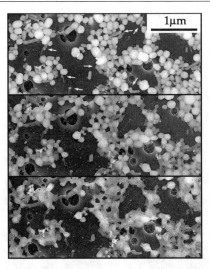

**Fig. 15** The images from *top* to *bottom* are of successively increased beam irradiation damaged TE-3170

while the pie-shaped domains would result from a domain remaining in crystalline register as the rod is bent back on itself. The blunted or club ends on many of the rods in Fig. 14b and d and Fig. 15 (arrows) would be consistent with a folding up of the rods during growth, but with much shorter fold segments.

Possible evidence for the folding, however, can be seen in Fig. 16a and b. Figure 16a is a high-resolution SEM micrograph of Cr-coated, air-dried sample G particles; it suggests a folded-rod morphology is present with a number of the particles, but with a rod diameter considerably less than half the particle diameter (as suggested by the diffraction contrast micrographs) and a number of folds. Several of these particles have a bent or blunted end (arrows). Further evidence for the multiple-fold model comes from the SEM micrograph in Fig. 16b, for which particles of TE-30 were compression-sheared before Cr-coating. Presumably the kinks in the rods correspond to the folds, the shorter distances between the kinks being similar to the particle lengths. Individual rods in this figure are several tenths up to approximately 2-μm long. The latter, corresponding to a molecular mass of approximately 750 000 Da if they contained extended chains, are much lower than the $\overline{M}_w$. Of interest is a comparison of the length of the rods and the chain length; the rods are on the order of 2 μm in resin G (Fig. 14a) and up to possibly 2 μm in the TE-30 sample in Fig. 16b, whereas the weight-average extended-chain lengths are considerably longer, 130 and 50 μm, respectively. We are thus left with the conclusion that the proposed mechanism of polymerization as rods (or ribbons) followed by their folding to form the particles is feasible, but both require folding of the molecules within the rods, significant molecular shear when the rods bend

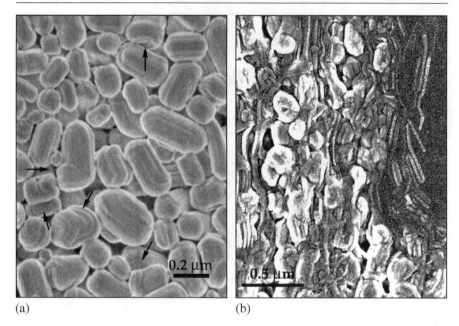

(a)                                                        (b)

**Fig. 16**   High-resolution scanning electron microscope micrographs of Cr-shadowed
(**a**) nascent resin G and (**b**) compression-sheared TE-30

back on themselves and then lateral correlation of the crystalline structure. To
our knowledge, none of these factors have been considered in discussions of
the polymerization process. Unfortunately, the polymerization conditions for
these standard-size resins are unknown; in a paper describing the structure of
particles of a nanoemulsion in which the polymerization process was stopped
at an early stage, Chanzy et al. [18] indicated a polymerization temperature of
80 °C in water.

### 3.1.2
### Nanoemulsion Particle Structure

Figure 17 shows unshadowed A18749 nanoemulsion particles. The particles
are essentially all rods, of various length up to approximately 1500 Å, have
a diameter from approximately 75 to 400 Å and often with pointed ends, They
presumably correspond to a stage of the polymerization before the surfactant
was depleted and particle folding took place. BFDC from a similar sample (inset) suggests that the rods are single crystals with the molecular axes parallel
to the long axis. 1500 Å corresponds to a degree of polymerization of 630, and
a molecular weight of 58 000. If the zero shear viscosity determined molecu-
lar weight ($\overline{M}_n = 200\,000$ Da, $\overline{M}_w = 49 \times 10^6$ Da) is accepted, or even anything
close, the chains must be folded at the ends of the rod particles. Indeed, with

**Fig. 17** Low-intensity image of Ausimont A18749 nanoemulsion. The *inset* is a diffraction contrast image of a similar nanoemulsion.

a diameter from approximately 160 to 240 Å in the longer rods, and a 4.9-Å spacing of the 100 planes in the hexagonal (phase IV) lattice, a rod of 200-Å diameter composed of a single, weight-average, folded molecule would have a length of 1200 Å, slightly longer than typical particles in the micrograph. We thus conclude all of the Ausimont particles consist of single or, at most, a few molecules, a conclusion of significant implication relative to the polymerization mechanism.

In contrast, the DuPont TE-5070 nanoemulsion particles are uniform in size and morphology (Fig. 18a), all being more or less hexagonal platelets approximately 700-Å thick (from shadow lengths) and 750 Å in diameter. ED (inset), in agreement with the shape, indicates the molecular axes are parallel to the thickness direction. BFDC (inset in the lower right in Fig. 18b)) and their shape indicate they are single crystals. Close packing of the particles during solvent evaporation appears to result in their deformation (Fig. 18b). Beam damage (left insets in Fig. 18b) results in particle shrinkage as shown by the enlargement of interparticle holes, the development of high density "points" (the white spots) and the formation of necks between touching particles. Although somewhat smaller in diameter, the particles closely resemble the oligomeric particles described by Chanzy et al. [18].

With $\overline{M}_n$ = 50 000 Da and a chain length of 1300 Å, the thickness is approximately twice the number-average molecular length, again suggesting chain folding, as single or multiple folds, in the as-polymerized particles. However, we know of no direct polymerization mechanism that would result in chain-folded nascent particles, of either the form here or in the rods of A18749. Although it is known that PTFE is insoluble in the polymerization medium, if

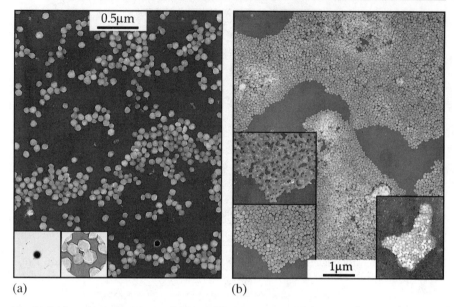

(a)                                                    (b)

**Fig. 18** (**a**) Low-intensity image of DuPont TE-5070 nanoemulsion particles. An ED pattern (100 reflections only) and the selected area used are shown in the *insets* (**b**) Densely packed particles. The effect of beam damage is shown in the *left insets*. The *right inset* is a bright-field diffraction contrast image.

the molecular weight determinations are accepted, the molecules for both nanoemulsions would either have to first grow to their final length in solution and then crystallize or have to reverse direction during addition of monomer to the ends of the growing chains.

Unfortunately, for attempts to understand the development of the particle morphology, polymerization history is not available for any of these samples, all being commercial products. Presumably the DuPont nanoemulsion is prepared similarly to the particles described by Chanzy et al., i.e., stopping a standard emulsion polymerization early, in their case at 1% conversion [18]. They suggest the molecules come out of the aqueous solution as very short oligomers (they indicate a polymerization temperature of 80 °C for "standard" resin), the crystals thickening by growth on the exposed molecular ends [17, 19], a situation unlikely to produce folded chains. If they were extended-chain crystals the "measured" molecular weight would be incorrect by a factor of approximately 2 and the distribution would be very narrow. In addition, if the observed morphology represents the initial stage of growth, with folded or extended chains, it is difficult to see how further growth would lead to the ribbons proposed by Rahl et al. [13]. Although they could continue to grow, as whiskers, to form fibrils or rods, most observed fibrils, including those described by Chanzy et al. [17, 19] in other papers at the same time, are smaller in diameter by a factor of 3–4.

A possibly similar feature occurs in the dilute, high-temperature solution polymerization of poly($p$-oxybenzoate) single-crystal whiskers [23]. Large-diameter, hexagonal, platelets grow at high monomer concentration. With decreasing initial concentration they appear to split as condensation polymerization continues, both within and at the ends of the particles. resulting in a coherent central core of a bundle of rods (whiskers). The individual whiskers, approximately 100 Å in diameter, which grow at low concentrations, are proposed to develop by the extended-chain crystallization from solution of short-chain oligomers which then undergo end-linking within the crystal lattice, giving off acetic acid, which aids the splitting of the central core. However, there is no evidence for a similar process here.

Desirable would be a study sampling a standard polymerization process as a function of time, a study most easily carried out in an industrial laboratory involved in PTFE polymerization. In the only such report we know of, Seguchi et al. [14] compared particles formed at 70 °C, 0% emulsifier, for 10 and 150 min, with the micrographs being similar to those in Fig. 11a. In both samples the relative population of particles and rods appears similar, with the rods and particles increasing in both diameter and length with polymerization time and the molecular mass increasing from 90 to $1300 \times 10^4$ Da. The Ausimont particles would appear to be formed under different conditions than the DuPont material. Although, on the basis of the study by Seguchi et al. [14], it would appear that a higher emulsifier concentration was used for the Ausimont sample, there must be further differences since both the rods (if extended chains) and the hexagonal crystals would be expected to start growth as thin single crystals of aggregated oligomers. Among the differences, it would appear that the Ausimont particles were initiated at various times, while the DuPont particles were initiated at the same time.

## 3.2
## Wide-Angle X-Ray Diffraction of Nascent Particles as a Function of Temperature

X-ray diffraction scans of a standard-size resin (dried sample G), TE-5069, a dry DuPont resin similar to TE-5070, and dried A18749 nanoemulsion are shown in Figs. ??–21. The results are useful for comparison with the effect of sintering the dispersion particles at similar temperatures. Considering first the standard-size resin, we show heating and cooling curves for sample G in Fig. ??a and b, respectively. The presence of 107 and 108 peaks at 25 °C is indicative of the hexagonal lattice, 15/7 helix, phase IV structure [24]. They disappear by 34 °C, owing to the "35 °C" transition to the phase I structure, merging to form the broad peak centered, e.g., at about 39° $2\theta$ at 100 °C. The 200 peak, buried as a shoulder of 107 at 25 °C, and the 110 peak shift to smaller angles with increasing temperature and nearly disappear at 375 °C. The peak labeled 300, on the other hand, appears to remain at a constant angle; we sug-

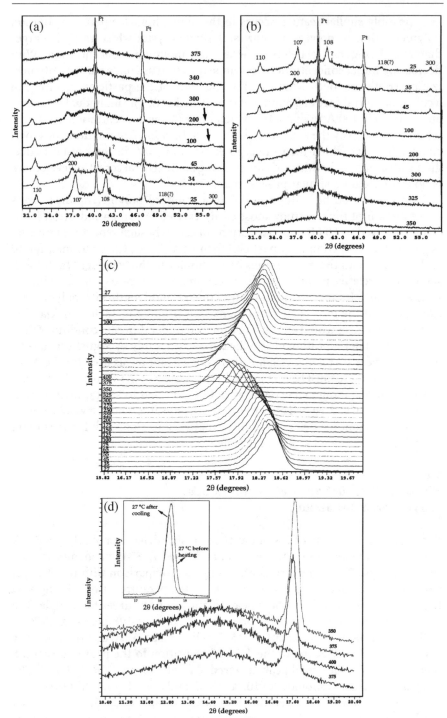

gest it is a compound peak of 300 from the PTFE and an inorganic impurity (possibly the platinum from the sample holder), the PTFE peak becoming separately visible owing to lattice expansion at about 100 °C (arrows). The peak at approximately 49.5° $2\theta$ is tentatively assigned to 118; however, it remains visible up to at least 100 °C, whereas the quadrant reflections in PTFE fiber patterns were shown to disappear at the 35 °C transition [24].

All of these effects were reversible with temperature (Fig. ??b), 110 and 200 reappearing when cooled to 325 °C, 300 at 200 °C, 118 at 100 °C and 107 and 108 below 35 °C. In a separate temperature scan of 100 (Fig. ??c) there was a similar decrease in scattering angle, the decrease increasing rapidly above 300 °C, with the peak broadening significantly at 350 °C and losing approximately half of its intensity between 350 and 375 °C. This figure is printed such that the broad, amorphous scattering background does not show. As shown in Fig. ??d, which shows only the high-temperature scans, at 400 °C the peak was absent. On cooling a much smaller peak reappeared at 375 °C, growing in intensity at 350 °C. It assumed its normal shape by 325 °C and increased in $2\theta$ with further cooling. On a second heating and cooling the curves in Fig. ??c and d were reproduced except that the intensity of the 100 peak at 375 °C on heating was considerably lower, i.e., more crystals had melted than when the nascent resin was heated, suggesting a greater degree of disorder and/or smaller crystal size. However, as shown in Fig. ??d (inset), 100 after melting is narrower and at a smaller angle than before melting, suggesting a larger lateral crystal size, corresponding to the development of the bands, and looser lateral packing; the greater melting at 375 °C is thus attributed to the less perfect packing in the crystals. In comparison with the high-temperature scans in the main part of Fig. ??d, almost no "amorphous scattering" (the broad peak centered at approximately 15°) is present at room temperature.

Elevated temperature scans of TE-5069-AN during heating are shown in Fig. 20a. At 27 °C the 107 and 108 reflections have already nearly disappeared, with 110 and 200 being visible up to 300 °C. An "amorphous" scattering develops at angles less than that of the 100 peak at temperatures above 300 °C, with a small 100 peak remaining at 350 °C (the maximum temperature for this scan, with 100 being similar, but relatively smaller than the 100 peak in sample G at 375 °C). Cooling the nascent resin below room temperature resulted in development of the 15/7 helix during cooling, as evidenced by the presence of the 107 and 108 peaks (Fig. 20b) which are retained down to at least 14 °C (the lowest temperature attainable by running cold water through the heating

◄  **Fig. 19** X-ray scans of resin G. (**a**) First heating, with temperatures indicated. Five minutes was allowed at each temperature for thermal equilibration for all samples. (**b**) Cooling. (**c**) 100 reflections as a function of temperature during heating and cooling. (**d**) 100 reflections at elevated temperature with the *lowest curve* being taken during heating, and the *upper curves* during cooling. 100 reflections at room temperature before and after heating to 400 °C are inset in **d**

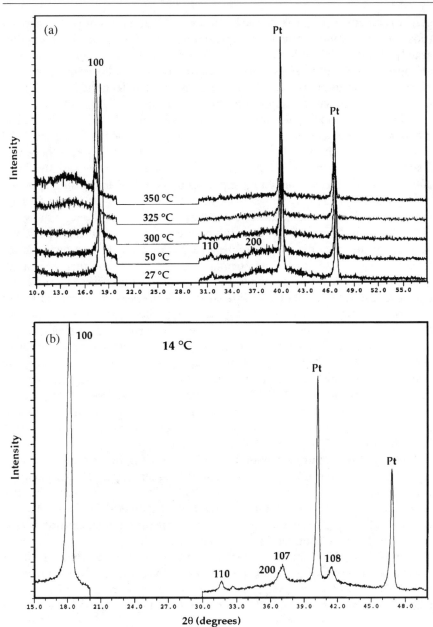

**Fig. 20** X-ray scans of TE-5069. (**a**) First heating, with temperatures indicated. The 100 peak is nearly gone at 350 °C. (**b**) Effect on the high-2θ region of cooling the nascent resin to 14 °C prior to any heat treatment. (**c**) Effect on the high-2θ region of cooling the resin to 14 °C after first heating to 400 °C. (**d**) Effect on the 100 reflection of cooling the resin to 14 °C after first heating to 400 °C. The *inset* shows the 100 reflection before and after heating to 400 °C.

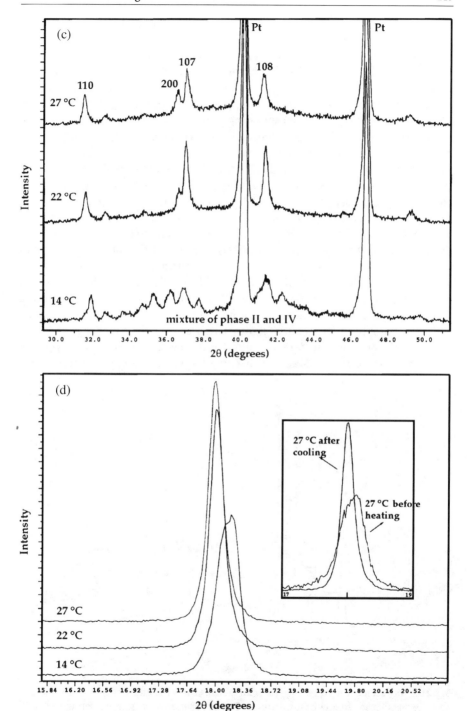

stage); this is below the "normal" 19 °C transition to a 13/6, phase II helix. After heating to 400 °C and cooling (Fig. 20c) the 107 and 108 peaks are present at 27 °C although they were not present initially in the nascent resin at 27 °C (Fig. 20a), and there is a partial transformation to phase II at 14 °C. As in the case of sample G, the 100 peak at 27 °C, before heating, is considerably broader and at a larger angle than after cooling (Fig. 20d). Cooling to 14 °C resulted in splitting of the 100 peak, corresponding to the presence of both phases. Also, as in the case of sample G, the 100 peak at 27 °C, before heating, is considerably broader and at a larger angle than after cooling (inset in Fig. 20d).

Retention of the hexagonal 15/7 helix structure during the first cooling of the nascent TE-5069 resin to 14 °C is attributed to the better packing of the chains in the as-polymerized lattice. This can also be seen in the low-temperature scans for the dried A18749 resin in Fig. 21; on first cooling the nascent resin, phase IV is present at 27 °C and is retained down to 14 °C. The 107 and 108 peaks, however, are relatively broad, possibly owing to the small cross-sectional area of the rods (Fig. 17) and are decreased in intensity and approached each other as the sample was heated to 27 °C. After melting at 400 °C and cooling, the 107 and 108 peaks are sharper and an amorphous scattering has developed at angles below that of the 100 peak. The 100 peak before heating was, again, broader and at a slightly larger angle before melting than after cooling. We attribute the X-ray results to more perfect crystals being present in the nascent resins, crystals that become larger in lateral dimensions after

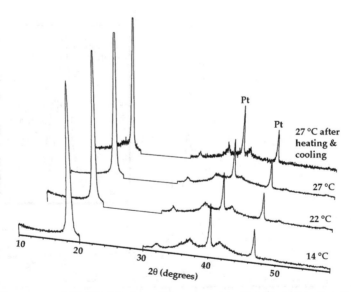

**Fig. 21** X-ray scans of A18749. The *lower three scans* show the effect of cooling the nascent resin to 14 °C prior to any heat treatment. The *uppermost scan* is after heating to 400 °C and cooling.

melting and recrystallization, but that are less perfect. As shown by Fig. 21, an amorphous component remains after the first heating to the "melt".

## 3.3
## Thin-Film Crystallization from the "Melt"

### 3.3.1
### Standard-Size Resins

For the standard-size DuPont dispersion particles, the effect of time and temperature in the "melt" was examined to a greater extent for TE-3170, with less extensive study of the other two resins. Figure 22 shows the effect of sintering time at 350 °C, with the samples in Fig. 22a–c slow-cooled by leaving them between the platens of the press as it cooled down. Five minutes of sintering resulted in no apparent change in particle size (approximately 2000-Å diameter) or shape (Fig. 22a). The only change was the smoother Pt/C coating as compared with its granulation on the virgin particles (lower inset) and the development of a thin film surrounding each aggregate of particles (upper insets). We suggest the film and the cause of the granulation is due to the emulsifier agent; the 5 min of heating is sufficient for it to flow off the particles, evaporating after 10 min of heating (Fig. 22b). Although granulation on the particles could be a result of aggregation of the Pt/C on the "liquidlike" surface, it does not occur on the films. This granulation is typical of the appearance of Pt/C or Pt/Pd shadowed nascent PTFE particles [4, 12] although it was not apparent on the Cr-coated particles used for the SEM micrograph in Fig. 16 or in Ref. [15]. The inset ED pattern in Fig. 22b, consisting of numerous 100 reflections, is from the selected area also inset. In the lower-left inset, from another area of this sample, some of the particles have developed "double striations", to be discussed further later. The arrows indicate small particles, considerably smaller then the nascent particles, in which the doubling has resulted in an M in the shadows, indicating the center between the striations is thinner than the striations themselves. The dimensions correspond to a molecular mass of approximately $11 \times 10^6$ Da.

Sintering at 350 °C for 30 min (Fig. 22c) results in the development of larger, angular particles. Several of the particles (large black arrows, more than 6000-Å long) have planar faces, giving the suggestion of the particles being hexagonal rods; several of the others (white arrows) have a planar face with steps suggesting the presence of thin lamellae parallel to the face. In addition to the particles, a thin film, either fractured more or less at random or at domain boundaries, presumably during cooling, covers the substrate. ED from this film (inset, from the region indicated by the small black arrow) is a single-crystal [001], PTFE pattern. If quenched into water after 30 min in the melt (upper-right inset in Fig. 22c) the structure resembles that obtained by slow cooling after 20 min of sintering time (not shown); the particle size is essentially unchanged although some merger, flattening of the boundaries and development of double striations (arrows) appears to have occurred.

Further enlargement of individual particles, up to nearly a micron in length, with most having one or more sets of double striations occurs after 60 min (Fig. 22d). "Tails" of varying length grew (inset in particular, from another area of the same sample); these consisted of a single double striation as shown more clearly in subsequent micrographs. Some granulation of the shadowing material has occurred, even on the substrate, in part owing to this sample having been overshadowed. After 2 h all evidence of the original particles disappeared with both lamellae and "mounds" with superimposed double striations (with an approximately 400-Å separation) present (Fig. 22e). We suggest the mounds are incipient bands; they appear similar to those described by Symon [9] for 2-h melt time at 380 °C (the only conditions described). Numerous small par-

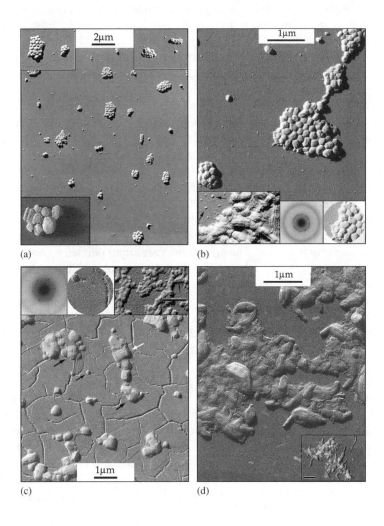

(a)                                          (b)

(c)                                          (d)

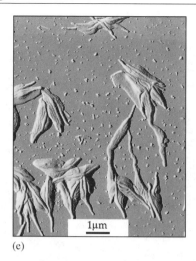

(e)

**Fig. 22** Morphology of dispersed TE-3170 held at 350 °C as a function of time prior to slow cooling: (**a**) 5 min; (**b**) 10 min; (**c**) 30 min; (**d**) 60 min; and (**e**) 120 min. The sample in the *upper-right inset* in **c** was quenched into water after 30 min in the melt with the *other insets* being from the same samples as the in the main micrograph. *Arrows* and *insets* are discussed in the text. In all images the *scale bars* in the *insets* are the same scale as in the main figure part, a nondescribed arrow indicates a region enlarged in the *inset*

ticles, with one dimension on the order of 1000 Å, are present on the substrate. Although not apparent in his micrograph, in other micrographs (e.g., the inset in Fig. 22b) they consist of short, double striations; we believe they are PTFE. With a molecular mass of 1.3 Da/Å$^3$ in the crystal, they clearly contain many more than one molecule, but, on the other hand, are smaller than the original nascent particles. As discussed further later, we suggest they are due to motion and aggregation of individual molecules on the substrate.

In TE-3698 held for 20 min at 350 °C and then slow-cooled (Fig. 23a), the structure resembles that for TE-3170 sintered for the same time; the original dispersion particles appear to merge to a substantial degree with the boundaries nearly disappearing. Double striations (approximately 400-Å spacing) are present on some of the elongated particles, the arrows indicating the regions enlarged in the insets. Further merger, with enlargement of the particles to more than 0.5 μm, occurs after 30 min (Fig. 23b). Thin, short "bands", resembling those previously described for fracture surfaces of PTFE melted at 380 °C [1–4] are present on several of the particles (arrows). Longer sintering times were not examined. The insets in Fig. 22c suggest the domains are single crystals.

Sintering TE-3170 at a higher temperature, 380 °C, for 5 min (Fig. 24a) results in some rearrangement and partial merger of the particles, including the development of double striations on some of them (inset). A thin film similar to that in Fig. 22a (350 °C, 5 min) surrounds each cluster of particles. The

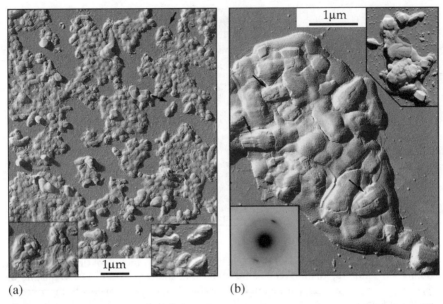

(a)                                                    (b)

**Fig. 23** Morphology of dispersed TE-3698 held at 350 °C for (**a**) 20 min and (**b**) 30 min before slow cooling. The ED pattern and the image with the selected area in **b** (*insets*) are from another negative.

rods in this micrograph are artifacts. There is further flattening, merger and development of larger particles, up to 0.75 μm, some with double striations, after 10 min at 380 °C (Fig. 24a, upper-left inset). Heating of the particles for 20 min followed by slow cooling results in bands (Fig. 24b) similar to those reported by Symons [9] for 2-h melt times. Double striations (550-Å spacing in the inset) extend the length of the bands, with short "tails", consisting of a single double striation, extending out from one or both ends. After 2 h in the melt, presumably for an area with fewer particles initially, the double-striation tails are longer with, in some cases, only the tail being seen (Fig. 24c). The striations are more clearly seen when the shadow direction is normal to the striation. For the same heat treatments, similar bands were observed for TE-3698 and TE-30, with the tails often being longer than a micron in TE-3698. With the original particles being relatively randomly dispersed, as in Figs. 14 and 15, there clearly is significant molecular motion during the structure formation, more so at 380 than at 350 °C. In comparison with the striations described by Wunderlich and Melilio [6] (Fig. 9) resulting from the presumed epitaxial rearrangement and folding of molecules on the surface, during annealing or cooling, of bands visible on fracture surfaces of Bunn et al. [1] type samples (Fig. 1), the striations shown here are much longer and more clearly defined; they resemble those described by Bunn et al. [1] (Fig. 2) and Symons [9] (Fig. 8).

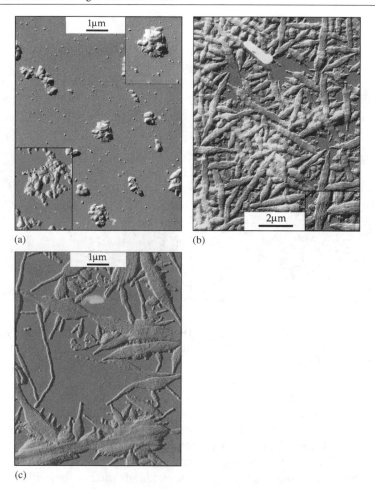

**Fig. 24** Morphology of dispersed TE-3170 held at 380 °C for (**a**) 5 min (*upper-left inset*, 10 min), (**b**) 20 min and (**c**) 120 min prior to slow cooling

Examples of the effects of sintering TE-30 and resin G are shown in Fig. 25a and b, respectively. Similar morphologies develop. Well-formed bands with superimposed double striations (450–600 Å spacing) are seen in the TE-30 sample after 30 min at 380 °C. As shown by the ED pattern from the selected area, the molecular axes lie in a plane normal to the band axis, likely parallel to the substrate. This would agree with the birefringence results of Symon [9]. In the upper-right inset in Fig. 25a, from a TE-30 sample sintered at 380 °C for a shorter time, (20 min on mica) isolated double striations (approximately 350-Å spacing) of varying length, up to approximately 1 μm, are present. Even the shortest particles are double striations, with ones such as that indicated by an arrow having a volume corresponding to 2–3 $\overline{M}_w$ molecules; many are much

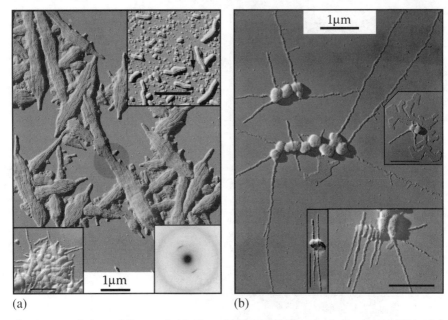

**Fig. 25** Morphology of dispersed (**a**) TE-30 held at 380 °C for 30 min (*inset* 380 °C/20 min) and (**b**) resin G held at 350 °C for 1 h (*main figure part* and *lower-right inset*) and 2 h (*center-right and lower-center insets*)

smaller. On the other hand, even the largest ones have a volume less than that of the original particles. We thus suggest, as emphasized later, that during sintering individual PTFE molecules diffuse away from the dispersion particles and can then be "trapped" on the substrate when cooled, crystallizing as single (or few) molecule, folded-chain single crystals. The larger ones, obviously, consist of a number of molecules, raising the question of whether they either formed in the melt or were a residue of an original particle. This is partially answered by the lower-left inset, of a similar 380 °C, 30 min sample quenched into water rather than slow-cooled. Although the bands are somewhat narrower, the double-striation tails are of similar size. As will be emphasized later, the structures observed are essentially independent of the cooling rate, suggesting they are forming in the melt. In the various parts of this micrograph none of the original particles can be identified.

In the main portion of Fig. 25b, of sintered resin G, individual double striations grow out from the dispersion particles that were held at 350 °C for 1 h. As shown in the lower-right inset, the rod particles may serve as epitaxial nuclei for the double-striations which then can continue growth on the substrate. The doubled nature of the striations is seen most clearly when they are on the particles. In the lower-center inset (380 °C, 2 h) two double striations are present on and extending from the particle; it is not known whether the particle is a short rod or an ellipsoid. At the center of the micrograph and in the center-right inset (380 °C, 2 h) folded-chain, faceted lamellae have grown parallel to the substrate.

In a recent paper Dürrschmidt and Hoffmann [25] describe their SEM and atomic force microscopy (AFM) studies of the effect of sintering single-layer films of relatively closely packed standard-size dispersion particles (product of Dyneon, approximately 2200-Å diameter) deposited by immersion and withdrawal of a "glass tissue" in a 60% concentration dispersion. They report band structures (labeled as worms), indicating their thickness (likely width, since it was from SEM images) was the same as the particle diameter and suggest they develop by interdigitation of extended-chain molecules in the nascent particles merging end-to-end. On the basis of our ED results from the double striations extending from the bands (Figs. 25a, 26d) and their likely relationship to the bands on fracture surfaces (Fig. 36b, c) and the birefringence results of Symon [9], the molecular orientation in the bands is perpendicular rather than parallel to their axis. In addition the width of the bands (and their thickness) can vary significantly, being considerably larger, e.g., than the A18749 particles whose sintering is discussed later. They also state a molecular mass of $7 \times 10^6$ Da, corresponding to an extended-chain length of 2.6 μm, an order of magnitude larger than their particle dimensions, rendering their model invalid. Of particular interest, however, is their Fig. 8, showing the effect of sintering at 400 and 420 °C for 5 h; this extends the temperature range we have used. Although the figure is not discussed in the paper, in addition to relatively narrow "worms" [possibly related to the on-edge lamellae or ribbons (double striations) in, e.g., Figs. 24c, 27a, 30] there are long, flat-appearing structures of unknown origin and structure; we have not observed similar structures.

### 3.3.2
### Nanoemulsions

### 3.3.2.1
### High Molecular Weight

With a lower melting point, similar structures were observed for TE-5070 and the Ausimont nanoemulsion particles for shorter melt times and/or lower temperatures. Figure 26 shows the effect of melt time at 350 °C on the resulting structures of the Ausimont resin. A melt time of 5 min (Fig. 26a) is sufficient to produce major changes in morphology. Many of the particles have merged to form larger structures, similar to those in Fig. 22e for TE-3170 after 120 min at 350 °C. They consist of a single lamella, approximately 400-Å thick, with a superimposed long double striation (approximately 400-Å spacing) and, in some cases, additional, shorter double striations. Numerous small, isolated particles are also present, as in Fig. 22e, here about 400-Å wide and of varying length. They thus appear larger in diameter than the nascent particles (approximately 200 Å, Fig. 17) but these are shadowed, with

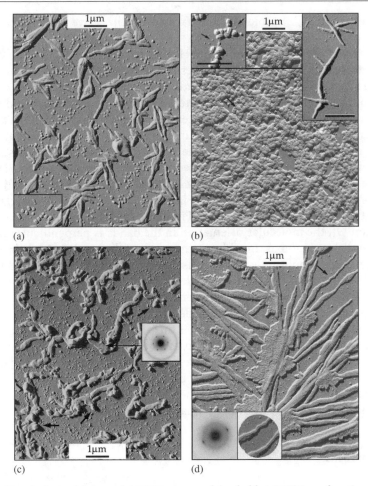

**Fig. 26** Morphology of dispersed A18749 nanoemulsion held at 350 °C as a function of time prior to slow cooling: (**a**) 5 min; (**b**) 10 min; (**c**) 30 min; (**d**) 60 min. *Black arrows* indicate regions in the *insets*; *white arrows* (in **d**) indicate highly retracted upper edges of lamellae

the shadowing contributing to the apparent diameter. At higher magnification nearly all of these particles consist of a single, short double striation (inset). For melt time of 10 min a number of different morphologies were observed in different areas of the same sample (Fig. 26b). In the main part of the figure all of the original particles have merged to form larger double-striation particles (approximately 500-Å spacing) of varying length, i.e., all wider than the original particles. The upper-middle inset is from near the center (arrow) of the large micrograph. The upper-left inset is from another region of the sample having a lower original density of particles; the arrows indicate very small particles that are doubled. The large upper-right inset is also from another area of this sample; it consists of single, long double striations with an approximately 300-Å spacing. Lamellae of the type in Fig. 26a were also present in this sample. The lack of differentiation between melt times of 5 and 10 min is likely due to the slow cooling in the press, some areas staying in the melt longer than others because of local variations in cooling rate.

For both 30 and 60 min (Fig. 26c, d, respectively) further merging occurs, with the (gradual) development of larger lamellae and elongated structures with, apparently, the double striations (inset in Fig. 26d). We suggest the elongated structures are related to the bands in, e.g., Figs. 22e and 24c and d. The inset diffraction pattern in Fig. 26c, obtained, in general, from the lamellae, indicates they are single crystals, with the chain axes normal to the lamellae. Of additional interest in this micrograph is the "excessive" height of the more or less cylindrical protrusions (arrows); they may be related to the whisker growths of TE-5070 in Fig. 35. The lamellae in Fig. 26c and d are quite thick, having a thickness on the order of 600 Å in Fig. 26d. However, with a number-average chain length of 5200 Å and a weight-average length of 0.127 mm the molecules are clearly folded. The ED pattern in Fig. 26d is from a region of single double striations; since only a pair of 100 reflections is present, it suggests the molecules are normal to the striation axis and either parallel to the substrate or at some angle to it. In none of this type of pattern was any other reflection (e.g., 107 or 108) seen that would help in determining the crystal orientation.

Of particular interest is the sample shown in Fig. 27a, heated at 350 °C for 30 min after dispersion on a mica substrate. Epitaxy was not evident on this sample but was observed for some other nanoemulsion samples sintered on mica (Fig. 27b) and on, presumably, NaCl (Fig. 27c). In Fig. 27a, in addition to the single double striation (approximately 250-Å spacing, 675-Å height) extending the entire width of the micrograph, even the smallest of the individual particles are double striations, with an overall width of approximately 450 Å. The lengths of many of them are also approximately the same, only the height varying. The apparent spacing of the long double striation is considerably smaller than in the particles, possibly because of the much greater thickness (height). For a "typical" particle of size $450 \times 800 \times 100$ Å (e.g., arrow), the molecular mass, if the particles are crystalline, would be approximately $50 \times 10^6$ Da, i.e., the particles might consist of a single, weight-average, folded molecule. The length and isolation from other material of the double striation in Fig. 27a, as well as the general appearance of the material in Fig. 26a, c and d, suggest considerable motion (flow) of the molecules on the substrate. We see no evidence, in the micrographs of the dispersed particles, for the linear array of particles that would be required for the double striation in Fig. 27 if motion did not occur. Even with the suggestion of motion, however, our only suggestion for the mechanism of the formation of such a linear structure with the corresponding high surface energy is that it is a folded-chain, on-edge lamella

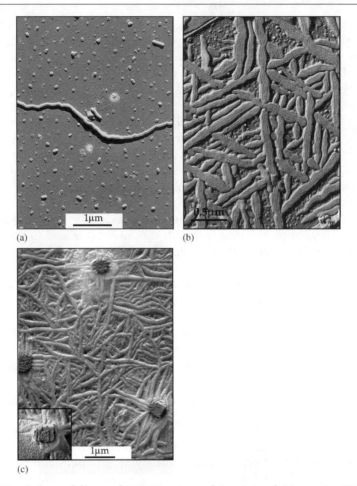

**Fig. 27** Morphology of dispersed A18749 nanoemulsion sintered (**a**) on mica (350 °C for 30 min) and (**c**) in the presence of adventitious NaCl on glass (350 °C, 2 h). (**b**) Epitaxially oriented bands of a nanoemulsion similar to A18749 sintered at 350 °C for 30 min on mica

(ribbon) which can only grow parallel to the substrate because of the difficulty of molecules to "climb" the fold surface to grow normal to the substrate.

As shown by Fig. 27c, NaCl appears to serve as a nucleating agent as well as an epitaxial substrate. Double striations are nucleated at the surface, growing out perpendicular, as well as being oriented on the surface parallel to the faces. In this case there even seems to be some climbing of the molecules up and over the NaCl surface.

Figure 28a and b contains micrographs of two samples of Ausimont nanoemulsion held at 380 °C for 20 and 30 min, respectively, before slow cooling in the press. In Fig. 28a the ends of large, well-developed bands are seen, each with numerous double striations; each has a total width of approximately

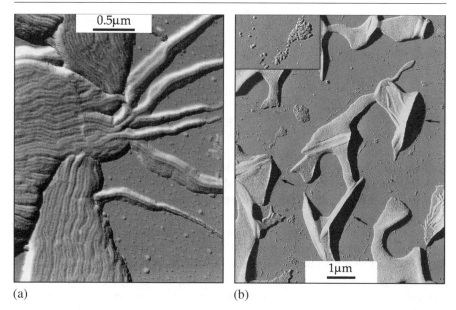

(a)                                        (b)

**Fig. 28** Morphology of dispersed A18749 nanoemulsion held at 380 °C for (**a**) 20 min and (**b**) 30 min before slow cooling

900 Å with a 350-Å spacing. As in Fig. 24d, for instance, several of the double striations extend individually from the ends of the bands, here for distances of several microns. Many have the appearance of growing up from a narrow lamellae lying parallel to the surface. The center-to-center separation of the two striations in the pair is smaller in the individual double striations (less than 200 Å) than on the bands. ED patterns of lamellae, as in Fig. 28b, are again 00l single-crystal patterns. Although the appearance of some regions of Fig. 28b (arrows) suggests beam damage may be causing shrinkage and pulling up off the substrate of the sample, this cannot be the case since the shadowing was done prior to insertion of the sample in the beam. The height of the shadows in Fig. 28b (arrows) suggests the lamellae are twisting up off of the substrate as they grow or cool. The small particles are about the same size as the smaller of the original particles; as shown in the inset in Fig. 28b, they often consist of short double striations. The arrow in the inset indicates a particle where the shadow is split, corresponding to the valley between the striations.

### 3.3.2.2
### Low Molecular Weight

A similar sintering time sequence for TE-5070 at 350 °C is shown in Fig. 29. In 5 min there is a smoothing of the boundaries, a decrease in thickness (to about 300 Å) and a corresponding increase in lateral size (to more than 1000 Å) of the original particles (Fig. 29a; a micrograph of shadowed, as-dispersed particles at the same magnification is shown in the inset). There is a further increase in size of the merged units (to more than 2000 Å) after 10 min (Fig. 29b). In some regions (e.g., inset) thin double striations (approximately 100-Å spacing) can be seen. As in Fig. 23b, the inset ED pattern and selected area, from another negative, suggest the domains are single crystals. A number of nascent particles must have merged to form each of the domains here. After 20 min lamellar spherulites were found in some areas with the boundary between two spherulites shown in Fig. 29c (for lower magnification of similar spherulites, see Fig. 29d). The overgrowth lamellae develop along splits in the basal lamella. On the left-hand side and the bottom of the figure the lower edge of the overgrowths are sloping, the upper, apparent growth face, edges are sharp; the opposite occurs at the upper left. Presumably the sloping edge corresponds to the growth face of the underlying lamella. In addition, however, there is also a ridge more or less along the center of each overgrowth that, in some cases, connects to the split in the basal lamellae (black arrows) and in other cases, in which they are less distinct, seems connected to the edge of a neighboring overgrowth (white arrows on the right). The white arrow on the left indicates an indistinct ridge for which there is no neighboring overgrowth. For all of the white arrows we suggest they indicate the edges of lamellae that lie underneath the surface lamellae. In a few cases the growth face of one of the overgrowths grows over another (black on white arrows). The latter is seen more frequently in the spherulites in Fig. 29d, of a sample sintered for 30 min The common orientation of the growth faces of the overgrowths is seen here also; in the upper part of the lower spherulite they all point in a counterclockwise direction, while in the upper spherulite they are in a clockwise direction. Also of interest is the texturing of the surface of the regions with a single lamella in both Fig. 29c and d, the overgrowths usually being smooth, and the thin "second edge" of the basal lamellae most clearly seen in Fig. 29c (white on black arrows). The texturing may be due to differential thermal shrinkage of the lamellae and the substrate, with granulation, on a scale larger than granulation of the shadowing material, being seen on many of the single-thickness lamellae in the various samples. ED from the approximately 250-Å-thick lamellae in the spherulitic samples are [001] patterns, again indicating chain folding, the number-average chain length being 1300 Å.

In the samples sintered for 20 min and longer, in addition to the spherulites, individual single crystals, single striations and bands with both single and double striations were also seen; the latter is shown in the inset in Fig. 29d, while the first two are shown in Fig. 30 (350 °C, 60 min) with appropriate ED patterns inset in Fig. 30b. In both of these micrographs, as for the basal lamellae in Fig. 29c, the single crystals have a double edge, the upper one appearing somewhat more rounded than the lower, but here, almost as thick. The striations appear single over most of both figures and again appear to grow up from more

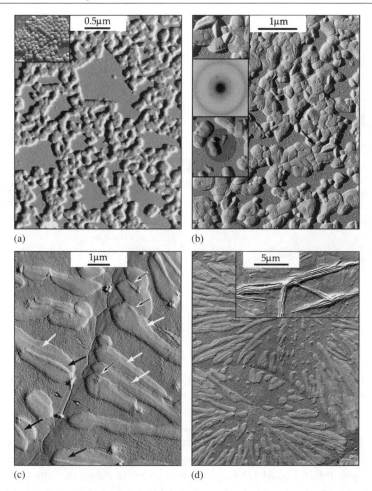

**Fig. 29** Morphology of dispersed TE-5070 held at 350 °C as a function of time prior to slow cooling: (**a**) 5 min (the *inset* shows shadowed nascent particles at the same magnification); (**b**) 10 min (the *upper inset* is from the area shown by the *arrow*; the *two lower insets* are ED patterns from a selected area from another negative); (**c**) 20 min; (**d**) 30 min. For a description of the *arrows*, see the text

or less ribbonlike lamellae lying parallel to the substrate. A possible normally spaced double striation is indicated by the arrow in Fig. 30a (inset). The substrates in Fig. 30 are "clean", indicating all PTFE has aggregated in the crystals and striations.

The effect of sintering time at 380 °C for TE-5070 is shown in Fig. 31. In 5 min, with slow cooling, single crystals and apparent single striations are already present (Fig. 31a). Overlapping lamellae, approximately 200–400 Å thick, are present in Fig. 31b (380 °C, 20 min); the granular regions are again believed to be a single lamella thick. For the overlapping to occur consider-

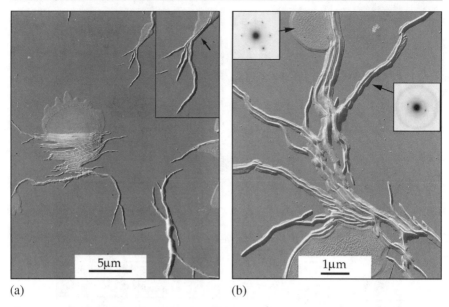

(a)                                     (b)

**Fig. 30** Morphology of dispersed TE-5070 held at 350 °C for 60 min. The *arrows* in **b** indicate the selected areas giving rise to the *inset* ED patterns

able fluidity and molecular motion must occur. ED from the striation regions again indicates molecular axes normal to the striations (approximately 200-Å spacing) and at some angle to the substrate (inset, Fig. 31c, 380 °C, 30 min), with ED from the lamellar regions indicating molecular axes normal to the substrate (inset, Fig. 31d, 380 °C, 30 min). Higher magnification (insets, from the regions indicated by the arrows) suggests the striations are doubled, with a small spacing. In Fig. 31d the overgrowth lamellae below the, apparently, inorganic square crystal are tilted with respect to the substrate; above the crystal they appear to become normal to the basal lamella, forming the striations. In contrast to the situation in Fig. 27b, the crystal here did not give rise to epitaxy. In nearly all cases the striations have an apparent inverted V cross-section, often appearing to grow up from a basal ribbon. In Fig. 31d the edges of the lamellae parallel to the substrate are doubled with the upper edge generally slightly retracted from the lower. In the regions indicated by the white arrows, however, the retraction is considerably larger, leaving an exposed single, lower lamella of thickness approximately 150 Å.

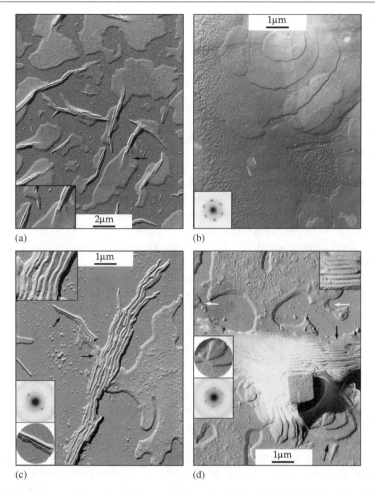

**Fig. 31** Morphology of dispersed TE-5070 held at 380 °C for (**a**) 5 min, (**b**) 20 min, (**c**) 30min and (**d**) 30 min

### 3.3.3
### Mechanism

The previous micrographs clearly show that the sintering time at 350 and 380 °C affects the resulting structure for both nanoemulsions and the standard-size dispersion particles; i.e., an equilibrium melt is not formed immediately, even at 380 °C. The nanoemulsions form "more perfect" structures for the same sintering time, with TE-5070 being more mobile than the Ausimont resin presumably because of its lower molecular weight. This would appear to be in contrast to linear polyethylene, for which reptation in the melt slightly above the melting point can lead to nearly instantaneous reentanglement and, presumably, random coil interpenetration across an interface (as in welding) [25]. With thicker samples one would have to consider the time required for the heat to diffuse into the sample; with the thin samples here that is not believed to be a significant effect, the platens having a large thermal mass. With the heating being done in air, degradation is a possibility; however, as indicated previously, we found no evidence of repolymerized PTFE, as described by Butenuth [11] and Melilio and Wunderlich [6] on cover slips covering the samples and the degradation rate constants at our sintering temperatures are very low [21, 22]. Another possibility is that PTFE does not "melt" at 350 °C to an isotropic melt but rather may be liquid-crystalline and, presumably, chain-folded in the thin films. With dispersed dispersion particles on a substrate, time would be required for molecular aggregation and motion across the substrate, domain enlargement and chain reorientation as the molecules in the dispersion particles "disengage", "wander" on the substrate and form the resulting larger morphologies. We have proposed similar chain folding in the mesomorphic melt for a random ter-polyester liquid-crystal polymer that crystallizes from thin nematic films and on surfaces of nematic melts as chain-folded lamellae [26, 27]. Indeed, there have also been recent suggestions, based on rheological and thermal analysis measurements, that even linear polyethylene has a liquid-crystalline character above the nominal melting point (up to 226 °C at atmospheric pressure) [28]. If so, the polyethylene "welding" would have to occur in the liquid crystalline state, with the "reentanglement" being a result of subsequent crystallization rather than random coil interpenetration.

In an attempt to see if there was structure in the melt, TE-5070 dispersions held at 350 °C for various times were quenched directly into water. After 5 min in the melt (Fig. 32a) substantial, but not total (inset), merger of clusters of the nascent particles occurred, yielding mounds often larger than 0.5 μm in diameter, with flat tops and slopping sides. Although likely not visible in the reproduction, many of the small particles in the background are short double striations; they are smaller than the nascent particles, having a volume ($6 \times 10^6$ Å$^3$) corresponding to approximately 160 molecules. In 10 min (Fig. 32b) there was further merger and molecular flow; the mounds have become thinner, although still tapered from a thicker, flat (now indented) center and flat lamellae have formed in some areas. In most regions there are discrete, visible boundaries between the domains; however, the arrow indicates a region where two of the flattened domains overlap and appear to have retained their identity. This raises the question of how the molecules in the overlying lamella got there, apparently having to "crawl" over the underlying lamella. After 20 min, only lamellar structures, approximately 100 Å thick, were found, resembling those observed for slow-cooled samples after 20 min sintering time (Fig. 29c) except that the edges are highly irregular and indented or "fingered" (Fig. 32c). In other regions striations were also observed. If the samples were thick enough well-developed spherulites and bundles of parallel striations were found (Fig. 32d, 350 °C, 60 min).

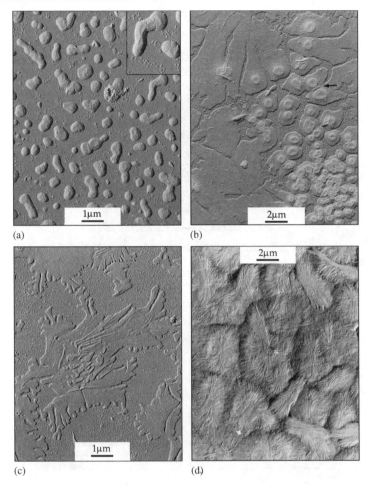

**Fig. 32** Morphology of dispersed TE-5070 quenched directly into water after being held in the melt at 350 °C for (**a**) 5 min, (**b**) 10 min, (**c**) 20 min and (**d**) 60 min

Water-quenching a A18749 sample after 30 min at 350 °C also results in irregular fingered lamellae as well as double striations (Fig. 33). We thus suggest that these films exist in the melt in the form observed after quenching; the fingering is attributed to poor wetting of the substrate at the melt temperature. Thus liquid crystallinity and chain folding in the melt, after sufficient sintering time, is suggested, with a high degree of molecular mobility on the substrate.

Further evidence for development of the folded-chain lamellae in the melt rather than by crystallization during cooling was obtained by reheating a previously well developed lamellar structure to 350 °C for an additional 30 min and then water-quenching it. As shown in Fig. 20, the lamellae remain well developed, with no sign of the fingered edges seen if quenched when first melted,

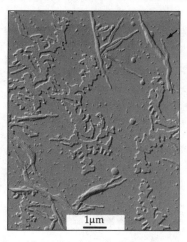

**Fig. 33** Morphology of dispersed A18749 quenched directly into water after being held in the melt at 350 °C for 30 min

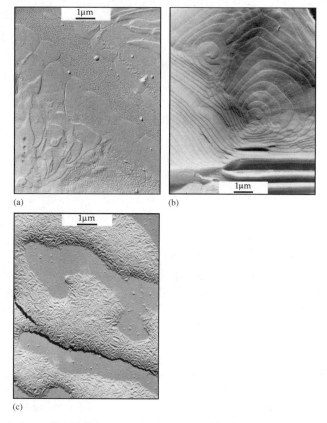

**Fig. 34** Morphology of (**a, b**) TE-5070 and **c** A18749 samples heated at 350 °C/30 min, slow-cooled and then reheated to 350 °C for 30 min followed by quenching into water

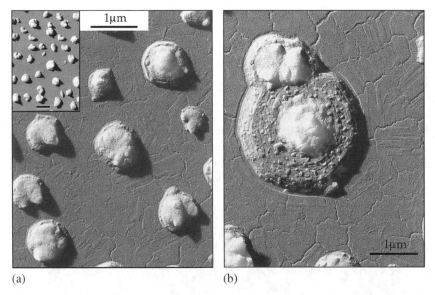

(a)                                                                                    (b)

**Fig. 35** Apparent PTFE whiskers grown from dispersed TE-5070 by holding in the 350 °C melt for 30 min followed by slow cooling

in both single-lamella-thick films (Fig. 34a) and in multiple-lamella-thickness films (Fig. 34b). The A18749 sample when treated similarly, on the other hand, changed considerably in surface texture (cf. Figs. 34c, 26d). In both cases double striations are seen, but in the reheated–quenched sample they are relatively short and superimposed on an underlying film; possibly this is because the sample in Fig. 34c is thicker than that in Fig. 26d.

Figure 35 depicts some unique PTFE structures found in one area of a 350 °C, 30 min, slow-cooled sample, the rest of which resembled Figs. 29c and d and 30. The shadowing in Fig. 35a indicates they are conical; although difficult to print because of their thickness, observation of the negative suggests the cones are built up of a stack of gradually decreasing diameter lamellae. Thus, these structures likely correspond to PTFE whiskers, the decreasing diameter being due to the limited supply of polymer from which they grow. With a height much greater than the nanoemulsion particle thickness, their growth again emphasizes the substantial molecular mobility of PTFE in the melt. The background in this figure resembles that in Fig. 22c of the TE-3170 resin. The inset ED pattern in Fig. 22c indicates it is PTFE rather than emulsifying agent; with a layer thickness of less than 100 Å it presumably consists of folded chains or segregated, very low molecular weight extended chains. A crack completely surrounds the whisker in Fig. 35b. None of the cracks are spanned by fibrils; although the film thickness corresponds to the order of ten folds per TE-5070 chain, the lack of fibers may not rule out chain folding depending on the fold plane orientation and the ease of lateral separation of the molecules.

## 3.4
### Bulk Samples of Nanoemulsion

## 3.4.1
### High Molecular Weight

Consideration is given to the sintering of the two nanoemulsions, samples of greatly different molecular weight. The morphological effects of sintering of standard-size resins have been discussed in the literature [1–4]. Figure 36a and b shows typical fracture surfaces of a bulk, compacted sample of A18749

(a)                          (b)

(c)

**Fig. 36** Fracture surfaces of thick, compressed A18749 resin sintered at 350 °C for (**a**) 5 min and (**b**) 30 min and (**c**) the sample in (**b**) reheated for an additional 30 min at 350 °C followed by slow cooling in a compression-molding press with no pressure applied

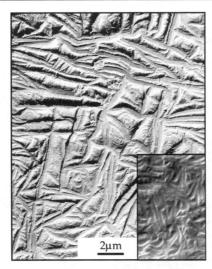

**Fig. 37** Free surfaces of the samples in Fig. 35. The *inset* shows the same sample prior to rinsing in methanol

sintered without pressure at 350 °C for (1) 5 min and (2) 30 min followed by slow cooling between the platens of the press used as an oven. The bands resemble those described in the literature previously for standard-size resins heated under similar conditions [1–4] except that the maximum thickness of the bands is only approximately 0.5 $\mu$m, corresponding to a molecular weight of approximately 200 000, i.e. $M_n$, but far less than the weight-average length of 127 $\mu$m. If the molecular weight measurements are accepted, the bands are, thus, chain-extended rather than extended-chain crystals. On the other hand, if the nascent rods are assumed to be composed of extended chains, typical chain lengths would be approximately one fifth of the striation lengths on the fracture surfaces. The pulling out of fibrils, with lengths up to at least 15 $\mu$m, during fracture, considerably longer than the length of the striations, suggests chain folding in the bands. As shown further later, the Pt/C granulates on the fibrils. Comparing the micrographs for 5 and 30 min of sintering, we find the bands in the 30-min sintered sample are longer and the thickest are somewhat thicker than in the 5-min sintered sample, again indicating sintering time as well as cooling conditions affect the morphology. Figure 36c is of the same sample as in Fig. 36b, reheated to 350 °C for an additional 30 min followed by slow cooling. Being at higher magnification, the bands and their striations are clearer, but no apparent change in structure occurred. In all three samples a wide variation in band thickness is seen.

"Free surfaces" of the same sample as in Fig. 36 are shown in Fig. 37, before (inset) and after rinsing the surface with methanol. The as-sintered surface is covered with indented regions that are similar in width to the bands on the fracture surface, but shorter. Rinsing with methanol apparently removes an ex-

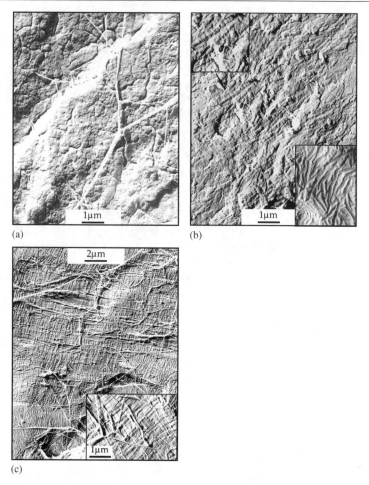

**Fig. 38** Fracture surfaces of quenched A18749 samples after sintering at 350 °C for (**a**) 10 min and (**b**) and (**c**) 30 min. A free surface of the 350 °C/30 min sample is shown in the *lower-left inset* in **b**

uded, low molecular weight coating, revealing the bands more clearly. There is even a suggestion of the presence of striations on their surface, oriented normal to the long axes on some of the micrographs. They thus differ from the bands in the thin films and also from the free surfaces of standard resins previously described. (Figs. 6, 7)

Quenching from 350 °C rather than slow cooling results in a more irregular structure. After 10 min of sintering, there is no appearance of bands, the particles apparently only merging to a degree (Fig. 38a) Thirty minutes of sintering at 350 °C, followed by water-quenching, leads to the development of thin (less than 0.1-micron) bands, in some regions relatively isolated in the merged particle matrix (Fig. 38b) and in others highly aligned over a substantial region (Fig. 38c). In the latter figure the striations on the bands extend across (are in register) a number of bands (inset). At the lower left, the bands appear tilted with respect to the surface, having the appearance of lamellae. It is noted that the fracture nearly always develops across the bands rather than parallel to their surfaces, similar to the situation for extended-chain polyethylene [5]. A free surface of the sample in Fig. 38b and c is shown in the inset in Fig. 38b; thin bands cover the surface and, as in Fig. 37, appear to be indented.[6]

### 3.4.2
### Low Molecular Weight

Attempts were made to prepare similar samples of TE-5069. Although TE-5069 has a tendency to flow out between the mold and platen when melted at 350 °C, thick samples were prepared by sintering in a well. When the thick samples were sintered at 330 °C for 20 min and slow-cooled (Fig. 39) or 350 °C for 5 min and quenched (inset), particle merger occurred in the interior, similar to that in Fig. 29a and b for short-time anneals at 350 °C. Five minutes of sintering, followed by slow cooling, on the other hand, resulted in the apparent development in the interior of small-diameter, thin, folded-chain lamellae (Fig. 40), while on the free surface linear features of unknown structure formed (Fig. 40, inset). A longer sintering time at 350 °C (15 min) resulted in the development of thin bands in the interior (Fig. 41a) with apparent merger across the end surfaces

---

[6] Rotating either figure by 180° results in hills appearing to become valleys and vice versa. The orientation of the print was chosen by observation of shadow directions on adventitious dirt particles and/or steep steps.

**Fig. 39** Fracture surface of a slow-cooled 330 °C/20 min TE-5069 sample. The *inset* is of a 350 °C, 5 min sample quenched in water

**Fig. 40** Fracture and (*inset*) free surfaces of a slow-cooled 350 °C/5 min TE-5069 sample

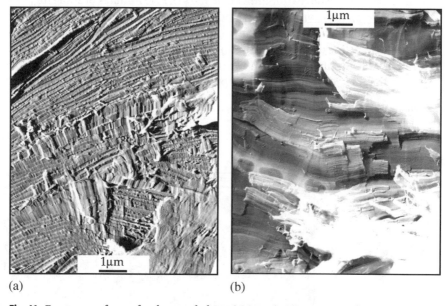

(a)                                                    (b)

**Fig. 41** Fracture surfaces of a slow-cooled 350 °C/15 min TE-5069 sample

occurring to various degrees (central region of Fig. 41a and all of Fig. 41b). In Fig. 41b the original bands are believed to have been oriented in the vertical direction. The thin bands, e.g., at the lower left of Fig. 41a are of a thickness approximately equal to the extended-chain length.

Sintering at 350 °C for 30 min also results in a varying fracture surface structure, varying from residual bands (Fig. 42a) to merged bands (Fig. 42b) to thin

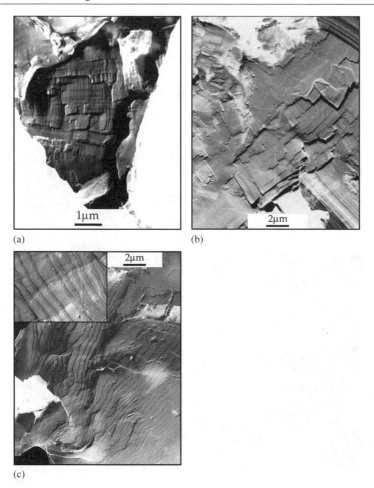

(a)    (b)    (c)

**Fig. 42** Fracture surfaces of 350 °C/30 min TE-5069 samples

lamellae. (Fig. 42c) Only the thinnest of the bands in Fig. 42a have a thickness corresponding to the extended-chain length; most are considerably thicker suggesting end-surface merger. We suggest the structure in Fig. 42b is due to nearly complete end-surface merger, with fracture between sheets of molecules in the merged bands. The linear steps would be due to steps in the fracture plane at the location of the original ends of the molecules. The thin lamellae in Fig. 42c we attribute to fracture at an interior void, i.e., the figure is of a free surface rather than a fracture surface. In some of the areas of this surface the lamellae have a double edge (inset), a feature also seen in several other melt-crystallized polymers [29] and still not explained.

Figure 43a–d shows micrographs of the free surfaces of the samples in Figs. 41 and 42. For sintering times of both 15 and 30 min lamellae of large lat-

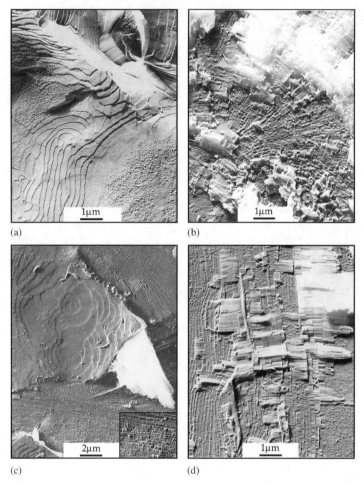

(a)                                           (b)

(c)                                           (d)

**Fig. 43**  Free surfaces of slow-cooled (**a, b**) 350 °C/15 min and (**c, d**) 350 °C/ 60 min TE-5069

eral extent, parallel to the surface, are found. In Fig. 43c they appear to be in a surface film that was partially removed during sample preparation, revealing thin (chain extended) bands underneath that are essentially perpendicular to the surface. In Fig. 43b the bands appear to radiate from a center, resembling a spherulitic structure; a similar structure was observed on the fracture surface of an extended-chain polyethylene sample (Fig. 3 [5]). As in the case of the fracture surface in Fig. 41a, the apparent chain-extended lamellae forming the bands merge in various regions to form thicker, more or less coherent bands. Unexpected, and unexplained, however, is why these free surfaces appear as they do, appearing the same as fracture surfaces. Since these are single-stage replicas, the shadowing Pt/C and backing C being removed from the original free surface, tearing of a surface film (Fig. 43c) and removal of a surface layer in Fig. 43b and d prior to shadowing should not have occurred.

Quenching of the sintered bulk samples after 30 min at 350 °C, as expected, yields fracture surfaces (Fig. 44a) resembling those for slow cooling for 15 min (Fig. 41b). The original bands, here, are presumably horizontal in the print, the end surfaces merging but retaining planes of easy fracture (arrow). Figure 44b is one of the few examples we found in which the fracture was apparently parallel to the end surfaces of the bands, the inset being from a region in which the fracture surface stepped down across several bands.

As in the case of the thin films, the observed morphology is primarily determined by the time at the sintering temperature, increased time (or higher temperature for the same time) leading to more complete rearrangement of the structure and formation of larger, more perfect, chain-extended (A18749) or extended-chain (TE-5069) lamellae (bands). Slow cooling results in a longer time at temperatures above the melting point than quenching and thus more perfect structures for the "same" sintering time. As for the thin films, we thus suggest the morphological structures observed are developed, in a "liquid-crystalline state" during the annealing or sintering. The increase in the thickness of the chain-extended crystals (bands) as a function of melt time is reminiscent of the crystallization of linear polyethylene from the "disordered hexagonal state" when crystallized under pressure; crystallization occurs by chain folding followed by chain extension while in the melt [30]. We suggest a similar

(a)                                                                    (b)

**Fig. 44** Fracture surfaces of air-quenched 350 °C/30 min TE-5069. The *inset* in **b** is from an area in which the fracture appears to have stepped up several bands

mode of structure development here, although likely from a nematic melt, i.e., without lateral order in the packing of the chain segments. As in the case of the dispersed particles on a substrate, time is required for the molecules to disengage from the particles and interact, forming first thin, folded-chain lamellae and these then undergoing chain extension. In thin films, and on surfaces of bulk samples, we propose the folded-chain structure, once formed, is retained. We have proposed similar chain folding in the nematic melt for a random ter-polyester liquid-crystal polymer that crystallizes, as indicated before, from thin nematic films and on surfaces of nematic melts as chain-folded lamellae, but with extended-chain crystals developing in the bulk with increasing time in the nematic state [26, 27]. We note there is no evidence for hexagonal packing at our sintering temperatures, phase I of PTFE, seen from approximately 35 °C to a melting temperature of 350 °C, resembling the disordered hexagonal phase of linear polyethylene in that the chains retain hexagonal lateral packing but undergo significant libration about their axes, helix reversals and a variability of helix pitch, approaching a planar zigzag [24, 31–33].

## 3.5
## Crystallization of and on Fibrillar PTFE

### 3.5.1
### Fibril Formation

The ability to draw approximately 100-Å diameter fibrils from PTFE dispersion particles, either by fracture after cold compaction [34] or by uniaxial or biaxial expansion of sheets or tubes after paste extrusion (extrusion of mixtures of the dispersion particles and lubricants such as mineral spirits) and lubricant removal [35], was demonstrated a number of years ago. The former process results in the development of a myriad of fibrils spanning the gap between the fracture faces; these were utilized for ED characterization of the PTFE conformation and crystal packing.

As shown in Fig. 45, for the nanoemulsion A18749, sufficient cohesion can be developed by simply drying a thick film of PTFE particles that fibrils will form spanning a crack developing during drying. This is despite the fact we assume these particles are each single molecules and therefore an individual fibril must involve cooperative unfolding and drawing out of molecules from several particles.

The latter process results in a network of fibrils spanning the gaps between nodes of undrawn material and is used for the production of a number of commercial products. Fibrils of presumably similar structure (Fig. 46) can be produced by large-scale shearing of dispersion particles dispersed on a substrate, e.g., by drying down a dilute dispersion on a glass slide and shearing it by drawing another slide across the particles. Dürrschmidt and Hoffmann [15] in their paper discussing the effect of sintering a layer of close-packed disper-

**Fig. 45** Coagulated film of A18749 that cracked on drying

sion particles show an SEM figure (Fig. 8b in [15]) in which nanofibrils span a crack in a film that had been sintered at 420 °C for 5 h. The sintered film consists of bands or on-edge lamellae similar to those shown in Fig. 12. Although they do not discuss the figure, we suggest the fibrils formed by differential shrinkage of the film and substrate during cooling, indicating the ease of fibril formation from sintered material as well as nascent particles.

Figure 46a–c is of a) C-coated, b) Cr-coated and c) Pt/C-shadowed samples of resin G. Similar fibrils are shown for TE-30, A18749 and TE-5070 in other micrographs later. For the standard-size resins the fibrils are of indefinite (TEM) length with neither the starting nor the ending point having been clearly identified. Both the carbon and Pt/C aggregate as isolated particles, with the Pt presumably being nanocrystals, on the suspended fibrils (as well as the particles) but not on fibrils lying on the substrate. These materials must be mobile on the fibrils, but whether this mobility occurs during the shadowing or during beam irradiation is not yet clear. In Fig. 46c the shadows of the fibrils are barely visible, in part owing to the shadowing direction. In Figs. 46a and 50b (below) the suspended fibrils appear to have uniform-width shadows, implying aggregation during beam irradiation. When the fibrils are metal-coated, the fibril itself appears to be visible, unexpectedly, as a more highly scattering structure in the core of the coating, with a diameter of less than 150 Å; when C-coated only, despite the greater similarity in electron density, the fibril core is not visible. The latter may be due to the thinness of the C coating. It is noted that the fibrils have sufficient strength to tear a surface film (inset) and lift the particles off the substrate, probably during beam irradiation, with nanofibrils being pulled up from the substrate (arrows in lower-left in-

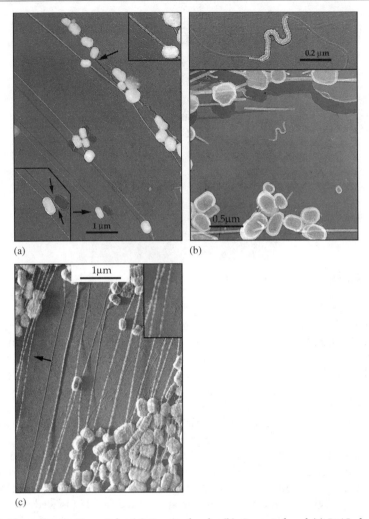

(a)                                                   (b)

(c)

**Fig. 46** Sheared resin G particles (**a**) C-coated only, (**b**) Cr-coated and (**c**) Pt/C-shadowed

set). The samples in Fig. 46a and c were annealed at 300 °C for 15 min and at 275 °C for 5 min, respectively; neither annealing condition affects the particles, but may result in a film of the emulsifying agent or low molecular weight PTFE being formed, the latter giving rise to the fibrils during lifting of the particles (Fig. 46a). The mechanism of development of the nanofibrils by shear is still unknown. They have no relationship to the finite-length rods, suggested to form the particles, that it was suggested can be seen in the compression-sheared sample in Fig. 16b; they are much smaller in diameter. While we have not attempted to observe fibrils in nonannealed, sheared A18749 and TE-5070 samples, as shown in Figs. 51b and 52b evidence for similar, long initial fibrils was seen in annealed, sheared samples as well as the nonsintered sample in Fig. 45.

## 3.5.2
## Effect of Sintering

The broadest range of annealing conditions was used for sheared resin G. Even though only part of the particles appear to have been affected by the shearing, annealing for only 5 min at 350 °C results in substantial rearrangement of the morphology. Slow cooling (Fig. 47a, b) results in shish-kebab formation on any fibril in contact with the substrate, with the kebabs consisting, in general, of single double striations. In addition there are isolated double-striation particles of varying length and, in Fig. 47a, a granular background; if the background particles are PTFE their volume (approximately $10^6$ Å$^3$) corresponds to a molecular weight of less than $1.5 \times 10^6$. On the other hand, the substrate in Fig. 47b, sintered under the same conditions, is clean. The suspended fibrils remain unaffected. In Fig. 47b, however, some of the shish have apparently retracted (arrows). Quenching the sample in water after 5-min annealing results in only incipient double-striation kebab formation and numerous irregular particles; (Fig. 47c); we suggest the latter are "wandering molecules" that were trapped on the surface individually or as "few molecule" aggregates during the quench.

Annealing the sheared samples for longer times at 350 °C can result in considerably wider kebabs, each again appearing, however, to have a single double striation on its upper surface (Fig. 48a). The kebabs resemble the bands in well-dispersed, annealed, undrawn samples in appearing to be wider on the substrate than on the top. This sample still has a granular background although other samples annealed similarly and for shorter times (Fig. 47b) had "clean" backgrounds other than relatively separated double-striation particles. In the inset the shish can still be seen between the separated kebabs. Air-quenching of the 30-min, 350 °C sample (Fig. 48b) resulted in narrower, more closely spaced kebabs and a background with small particles that were more isolated, but whether the difference is due to the cooling rate or variations in the initial particle concentration is not known. The BFDC in the inset is discussed later. Further evidence for the presence of a PTFE film on the substrate before cooling (Fig. 32) was observed in a water-quenched, 350 °C, 30-min-annealed sample (Fig. 48c). The double striations at both ends of the lower-right arrow appear to have grown on the film, oriented normal to the apparently underlying shish at the tip and randomly at the tail of the arrow. The kebabs are not much wider than in the water-quenched 5-min annealed sample in Fig. 47b. Annealing for 10 min at 375 °C results in fingerlike double-striation outgrowths from undistorted nascent particles, and closely spaced kebabs, but still leaves the suspended fibrils unaffected (Fig. 49).

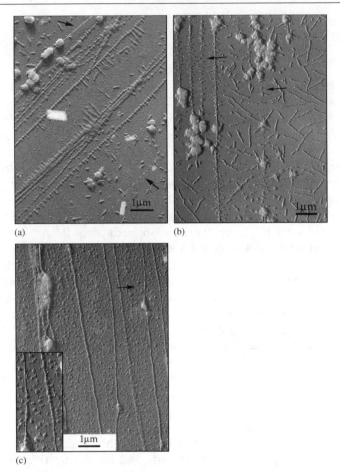

**Fig. 47** Sheared resin G particles sintered at 350 °C for 5 min and (**a**) and (**b**) slow-cooled or (**c**) quenched in water

Similar shish-kebabs were also grown by sintering sheared TE-30 (Fig. 50). Again the spacing and width of the kebabs has no relationship with the time of annealing; rather we suggest it is related to the amount of nearby material from which it "draws" molecules during growth. In Fig. 50a the ke-bab spacing in the insets is about 1500 Å, consisting of single double striations that are only slightly wider at their base than at the top. In Fig. 50b, for a simi-larly treated sample of TE-30, a number of the original particles remain, with both suspended fibrils with widely separated Pt beads (black arrows), short shish on the shish-kebabs where the fibril was in contact with the substrate and double-striation outgrowths from and on the particles (white arrows).

After 10 min of sintering (Fig. 50c) many of the kebabs contain several dou-ble striations and are more randomly oriented relative to the shish core (which,

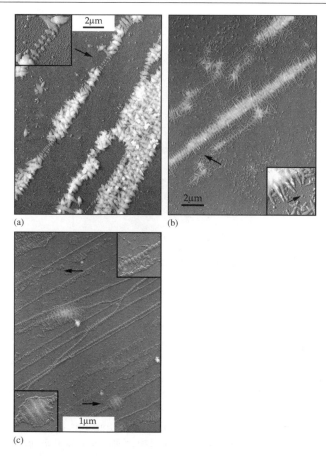

**Fig. 48** Sheared resin G particles sintered at 350 °C for 30 min and (**a**) slow-cooled, (**b**) air-quenched and (**c**) water-quenched

in general, has retracted), with similar bands also grown from the dispersed particles. Air-cooling after 20 min of sintering at 350 °C results in similar kebab bands and randomly oriented bands, as well as isolated multiple molecule, double-striation particles (Fig. 50d).

The insets in Fig. 50c are excellent examples of BFDC from the kebabs (bands); the diffracting regions appear to consist of lines about the width of a single striation with crystal registry extending diagonally from striation to striation in places. This would appear to contradict our suggestion that the double striations correspond to a single folded-chain lamella, the striations being the Pt nucleated on the folds; possibly the diffraction contrast image is shifted relative to the shadowing image. This type of image is discussed further later, relative to similar observations for sheared, annealed A18749 and TE-5070.

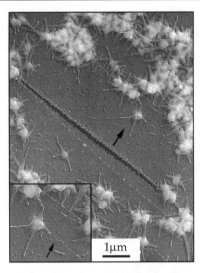

**Fig. 49** Sheared resin G particles sintered at 375 °C for 10 min

In limited observations, fibrils were not observed in sheared A18749 samples; that they could form is shown in Fig. 45, where they span a crack in a coalesced film of nascent particles. Sintering of sheared films at 350 °C for 30 min followed by slow cooling, produced rows of bands similar to those of TE-30 in Fig. 50c (Fig. 51a); as in that figure BFDC consists of lines approximately the width of a single striation which again lie at an angle to the striations. The contrast extending across more than one double striation suggests lattice registry of neighboring on-edge ribbons or lamellae, if that is what comprises the bands. On the basis of the high strength of the contrast, we assume it is due to 100 planes being in diffracting orientation. Although the narrowness of the diagonal lines could be due to rotation of the lattice about the molecular axis, we have no explanation as to how the lattice orientation propagates across the presumed fold surface to give rise to the diagonal lines. Rather, we tentatively suggest the contrast is due to an underlying lamella lying parallel to the surface; i.e., the broad base shown in Figs 22, 26a, 28a, 30b and 31b and c. An individual, flat-on lamella is present in the lower right with several sharp diffraction contrast lines; see also Fig. 52c later.

No shish are present in Fig. 51a. Indeed there is some question as to whether the oriented rows of bands are shish-kebabs; many of the kebab bands lie at an oblique angle to the apparent shear direction. Possibly they were nucleated along lines of smeared particles. On the other hand, Fig. 51b shows unmistakable shish kebabs (350 °C, 30 min, followed by air-quenching). In the shish on the left all of the kebabs would be nucleated at the sides of the apparent fiber, a split being present between the kebabs growing in opposite directions. On the right shish, however, a number of the kebabs extend across the cen-

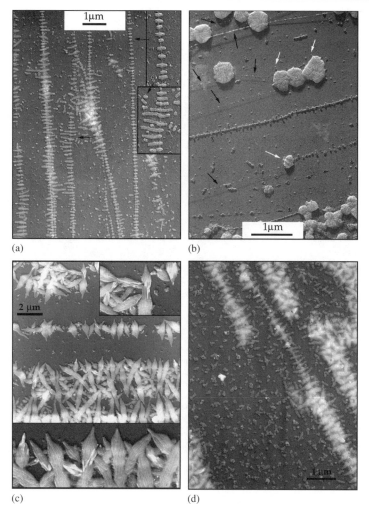

(a)                               (b)

(c)                               (d)

**Fig. 50**  Sheared TE-30 dispersion particles sintered at (**a**) and (**b**) 350 °C for 5 min, (**c**) 350 °C for 10 min, all slow-cooled, and (**d**) 350 °C for 20 min and quenched in air

ter line, possibly due to nucleation on the top of the underlying fibril, with the nucleated lamella then growing in both directions. Interestingly, and without explanation at present, the on-edge double-striation ribbons all curl, apparently randomly relative to direction. The kebabs consist of both short and long ribbons, with additional curled ribbons forming the matrix; some of the ribbons make complete circles. Similar curling was also observed for nonsheared samples (Fig. 51c). The curved bands appear to develop from lamellae lying at an angle to the surface; in the inset there is a suggestion they are double-edged, as in Fig. 32c. On the basis of discussions at PP' 2004 (Dali) relative to this figure and those in Ref. [36], it is suggested that they are related to the twisted lamellae

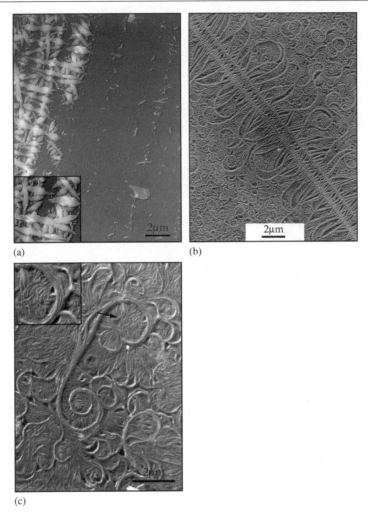

**Fig. 51** Sheared A18749 sintered at (**a**) 350 °C for 30 min and slow-cooled and (**b**) 350 °C for 30 min and air-quenched. A nonsheared sample is shown in **c**, also sintered at 350 °C for 30 min and slow-cooled

in banded spherulites (see Fig. IV-64 in [4]), curling instead of twisting when the film is thin; possibly thin enough that twisting would require twisting out of the thickness of the film.

Annealing of sheared TE-5070 also gives rise to apparent "shishless" shish kebabs (Fig. 52a), here sheared and annealed at 320 °C for 5 min. The low molecular weight of TE-5070 results in both a lower melting temperature and, apparently, retraction of the fibrillar material during the sintering. As shown in the inset, the kebabs consist of a short, single double striation superimposed on a nearly circular or elliptical mound, as do all of the individual particles

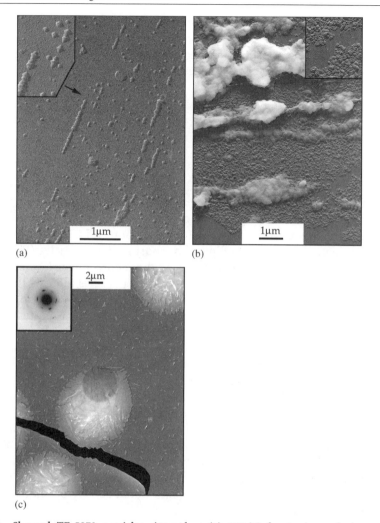

(a)                                                        (b)

(c)

**Fig. 52** Sheared TE-5070 particles sintered at (**a**) 320 °C for 5 min, and slow-cooled (**b**) 350 °C for 30 min, and water-quenched and (**c**) lamellae in bright-field diffraction contrast from a 350 °C/30 min, air-quenched sample. The *inset* ED pattern in **c** is from the overexposed selected area.

as well. That fibrils can form, despite the low molecular weight, is shown in Fig. 52b, of a TE-5070 sample sheared and sintered at 350 °C for 5 min and water-quenched. Suspended fibrils remain after the sintering. In addition, as shown in the inset, nanofibrils were drawn out across the gap between the particles, apparently as the sample cooled. The individual particles, consisting of short double striations, have a volume of approximately $5 \times 10^6$ Å$^3$, corresponding to a molecular mass of approximately $6 \times 10^6$ Da, much larger than the molecular mass of the TE-5070. In other regions of similar samples, large flat-on lamellae formed, with numerous sharp lines of diffraction contrast (Fig. 52c). Again, we have no explanation for the sharpness of these lines since one would expect reasonably large regions in the (presumed) single-crystal lamellae to be in a diffracting position even if the lines are due to bend contrast.

# 4
# Conclusions

1. Nascent particle morphology
    (a) The morphology of the standard-size resins and the mechanism of their formation are still unclear. There is reasonable evidence that the initial stages of their polymerization involves the formation of extended-chain fibrils or rods (Figs. 11, 16b) or ribbons (Fig. 12) that fold back on themselves to form elliptical or short rodlike particles (Fig. 14) as the emulsifying agent is used up by the increasing surface area. The diffraction contrast images from the final particles indicate the width of the coherent diffraction regions is equal to the radius of the particle, even at the ends of the particles, and sometimes the particle diameter (Figs. 13, 14). Thus, the particle diameter would involve, at most, two initial rods or the lattices from neighboring folded rods or fibrils would have to become in-register. This also requires that the lattice remain (or redevelop) coherent as the rod is bent back on itself at the end of the particle. In dispersions containing rod and ellipsoidal particles, the rod diameter is often approximately half of the particle diameter (Fig. 14a–c). The blunting of the ends of some of the rods suggests a folding from one end. Both of the second possibilities for the diffraction contrast images require only rotational molecular motion at room temperature, the type of motion proposed for the 19 and 35 °C transitions [24], since the molecular axes would already be aligned in the appropriate direction.
    (b) The morphology and size of the A18749 nanoemulsion (Fig. 17) is consistent with an individual particle consisting of a single molecule, with the molecular axis parallel to the rod axis and *folded* at the ends. Even if the molecular weight measurements are in error to some extent, the particles must consist of only a few, *folded* molecules. We know of no polymerization mechanism that would lead to a folded chain. The PTFE molecules presumably come out of solution as short-chain oligomers; rod crystals could form by their aggregation and axial alignment, but they would then be expected to grow as extended-chain filaments or rods depending on the number of oligomers that aggregate laterally. Diffraction contrast suggests the rods are single crystals (Figs. 13b, 17).

(c) Although the difference in the molecular-weight-predicted extended-chain length and particle thickness for the TE-5070 particles is less than for the A18749 ones, the number-average extended-chain length is approximately twice the particle thickness, again indicating chain folding (for A18749 the comparison was made for the weight average). Diffraction contrast and the hexagonal shape suggest the particles are single crystals (Fig. 18). If they were folded or extended-chain structures, of interest would be the distribution of the molecular weight and the location of the chain ends.

2. X-ray diffraction vs. temperature

A comparison of X-ray scans before and after sintering at 350 °C indicates a larger lateral crystal size, looser molecular packing and a reduced crystallinity for both the standard-size resin and the nanoemulsions Fig. 21. Both the transition (19 and 35 °C) temperatures and the disappearance of lateral ordering (100) for the low molecular weight TE-5069 PTFE occur at a lower temperature than for the nascent nanoemulsion A18749 and the standard-size resin G. "Melting and recrystallizing" resulted in an increase in the transition temperatures.

3. Sintering of dispersed particles

(a) The morphology of dispersed standard-size dispersion particles, after cooling, is a function of the time and temperature at the sintering temperature, a temperature sufficiently high that lateral order is lost as demonstrated by the X-ray diffraction scans. Short-time sintering at 350 °C results in, possibly, flow of the emulsifying agent to the substrate (Figs. 22a, 24a), followed by its apparent evaporation, as well as particle merger (Figs. 22a, d, 23a, b, 29a, b). The merged particles often appear faceted, possibly hexagonal (Figs. 22c, 23b) After extended times at 350 °C (Fig. 22d, e), or shorter times at 380 °C (Fig. 24a and inset) a rearrangement of the morphology occurs with folded-chain lamellar crystals developing (Fig. 22e) parallel to the substrate and, apparently, as double-striation on-edge ribbons (Figs. 22d, e, 24b, c). ED indicates the molecular axes are normal to the double striations (Fig. 26d) and thus are chain-folded. Bands also develop, with several double striations on their surface, often with one of the double striations extending out from the end of the band over the substrate (Figs. 22e, 24b, c). Not yet resolved is whether the double striations on the bands are inherent to their structure (i.e., they consist of several on-edge lamellae) or they develop owing to folded-chain, epitaxial crystallization of PTFE molecules on underlying chain-extended crystals as is proposed for the structure of bands in bulk samples (see later). For short times of sintering double striations develop on the surface of the merged particles parallel to their longer axis, (Figs. 22b, 22c, inset, 23a); although again possibly owing to epitaxial crystallization of "exuding" PTFE molecules, if the attributions of the double striations to folded-chain, on edge ribbons and the interpretation of the faceting are correct one might expect the orientation of the double striations would be parallel to the short axis. In addition, at intermediate sintering times, individual molecules appear to disengage from the particles and "wander" on the substrate, forming single-molecule or few-molecule folded-chain, double striation, on-edge lamellar single crystals (Figs. 22e, 24a, 25a, 26a, b, 27a, 28b, 47a–c, 48b, 49, 50a–d, 51a).

(b) Sintering of nanoemulsion A18749 at 350 °C results in similar morphology changes. The rate is faster, possibly because of the larger surface area. At 350 °C lamellar folded-chain crystals parallel to the substrate (Fig. 26a, c, d), merged particles with superimposed double striations (Fig. 26b), bands with both superimposed and outgrowing double striations (Fig. 26d) and isolated single-molecule or few-molecule single crystals (26a–c) were seen. In some samples (Fig. 26a) a single double striation was found on an apparent lamellar folded-chain single crystal, suggesting epitaxial crystallization on the fold surface. NaCl (Fig. 27c) and possibly mica, can cause epitaxial crystallization of PTFE, the molecules lying on the substrate surface. Quenching of 350 °C/30 min sintered A18749 into water resulted in folded-chain lamellae, with irregular outlines, lying parallel to the substrate (Fig. 33).

(c) Sintering of TE-5070 at 350 °C also results in particle merger (Fig. 29a, b), with superposition of double striations (Fig. 29b) followed by large folded-chain, lamellar single crystal and spherulite growth (Fig. 29c, d). Structures possibly related to bands, consisting apparently of on-edge, in this case single–striation ribbons were found for extended sintering times (Fig. 30). The lamellae parallel to the substrate in these samples had a double edge, a feature also seen in several other samples of both TE-5070 and other resins (Figs. 31a, d, 42c). At 380 °C lamellar crystallization was much rapider (Fig. 31). Double striations were observed for these samples, but much more closely spaced than in the higher molecular weight samples; in Fig. 30 they may have been too closely spaced to resolve. Five minutes of sintering at 350 °C followed by quenching into water resulted in merger of most of the particles into "mounds" with flat tops and slopping sides. Small double-striation particles were also seen, interpreted as clusters of molecules trapped on the substrate during the quench; they are smaller than the original particles (Fig. 32a). After 10 min of sintering, the mounds have become thinner, flat tops remain, and a substantial portion of the material is in the form of folded-chain, irregular-outline, lamellar crystals (Fig. 32b). After 20 min of sintering the mounds had completely converted to lamellar crystals, with small particles still present in the "open" areas (Fig. 32c); in thicker regions the lamellar spherulites were formed. (Fig. 32d).

The double-striations in TE-5070 samples (Figs. 30, 31a) and A18749 (Fig. 28a) appear to grow up from (or on) parallel narrow lamellae.

(d) A sintering time and temperature dependence of the morphology was observed for the three types of samples, dispersions with standard-size particles and both high and low molecular weight nanoemulsions. It is suggested that the folded-chain morphology develops with time in the "melt", the melt at longer times (and/or higher temperatures) consisting of aligned molecular segments. A similar suggestion has been made for a ter-polymer liquid-crystalline polymer; when crystallized in thin films or on surfaces of bulk samples chain-folded lamellae form and remain, whereas in the interior extended-chain or chain-extended crystals formed [26], with the chain-folding being attributed to surface effects and/or the thinness of the films from which they were crystallized [29]. Once formed, the well-formed lamellar structure was retained if a sintered TE-5070 sample was reheated to 350 °C and water-quenched (Fig. 34a, b), whereas A18749 underwent an as yet uninterpreted change in morphology (Fig. 34c).

(e) The results require that substantial molecular motion, either as individual molecules or as small clusters, occurs on the substrate during sintering. It is not believed that the structures observed were the result of depolymerization and subsequent repolymerization, as described by Butenuth [11] and Melilio and Wunderlich [6]; no material was found on cover slips covering the samples during sintering. In several of the micrographs, particularly of the whiskers (Fig. 35a, b), there was evidence that the PTFE molecules could "climb" some distance off the substrate to form lamellar crystals on top of each other. Of particular interest are the single-molecule or few-molecule single crystals observed for many of the sintering conditions and samples, in particular for the samples with high molecular weights. Under suitable sample preparation conditions, molecular weight measurements by TEM or AFM might be possible.

The double-striation morphology has tentatively been attributed to Pt nucleation on the fold edges of on-edge lamellae, similar to Au nucleation on the fold edges or interlamellar regions of, e.g., on-edge lamellae in drawn, annealed films of polyoxymethylene [37] and epitaxially crystallized samples of various other polymers [38]. AFM is currently being used to characterize further this feature.

4. Sintering of coagulated nanoemulsion particles

(a) Sintering of bulk A18749 at 350 °C, followed by slow cooling resulted in a morphology similar to that previously described (Figs. 1, 2 as compared with 26a–c). Although the sintering times for Figs. 1 and 2 were not given, other samples with similar resulting structures were sintered for 2 h before slow cooling. Thus, considerably shorter sintering times were apparently needed to obtain the chain-extended bands in A18749, 5 min plus slow cooling for Fig. 36a. When sintered samples were quenched, the development of the band structure in the melt could be followed, the particles initially merging (Fig. 38a), with thin bands then forming which gradually thicken with sintering time (Fig. 38b, c). Thus, again the results need to be interpreted in terms of the structures developing in the melt. Free surfaces of sintered bulk A18749 appeared to consist of thin bands in both slow-cooled and quenched samples; lamellae were not seen. (Figs. 37, inset, 38b, inset)

(b) Short-time sintering of bulk TE-5069 (coagulated TE-5070), at temperatures as low as 330 °C, also resulted in particle merger (Fig. 29), with thin, apparently folded-chain lamellae then developing at 350 °C followed by chain extension. The latter results are similar to the suggested mechanism of formation of extended-chain crystals in polyethylene crystallized under pressure [30], chain folding occurring in a "disordered hexagonal phase" in which there is sufficient chain mobility that chain extension then occurs, the extended-chain (or chain-extended) crystals growing in the melt phase. In TE-5069 this is followed by an apparent merging across the end surfaces of the extended-chain crystals (Figs. 41, 42a, b). Unexpectedly, and to date unexplained, free surfaces of the sintered TE-5069 had similar chain-extended and merged chain-extended bands (Fig. 43b, d). Thin lamellae were also observed (Fig. 43a, c). Quenching of the samples from the sintering temperature resulted in structures similar to those obtained by sintering for shorter times, followed by slow cooling (Fig. 44).

(c) The results for both A18749 and TE-5069 are again interpreted in terms of the observed structures developing while the samples are in the melt. Despite the large chain length of the A18749 (weight-average chain length is approximately 0.13 mm), it is suggested the single-crystal particles first merge, a feature apparently eased by the initial chain alignment, with the molecules then unfolding from the particles, possibly forming folded-chain lamellae (A18749) and then undergoing chain extension. The mechanism of the end merging of the bands in TE-5069 is not known; possibly chain extension of a chemical nature is occurring by "end-linking". Molecular weight measurements after sintering would be useful.

5. Sintering of shear-deformed dispersion particles

(a) Shearing of both standard-size dispersion particles and both types of nanoemulsion particles yields nanofibrils of approximately 100-Å diameter and indefinite length (Figs. 45, 46, 52b). To date we have not been able to identify, with certainty, the ends of the fibrils or the mechanism by which they are formed. Of particular interest is the ability of TE-5070 to form the fibrils despite its low molecular weight.

(b) Sintering of suspended nanofibrils, for temperatures up to 375 °C (Fig. 49) and times as long as 2 h (not shown) produced no visible change. On the other hand, nanofibrils lying on the substrate served as nuclei for molecules "wandering" on the substrate, resulting in formation of shish kebabs (Figs. 47–52). The kebabs appear to be identical to the double-striation on-edge ribbons growing out from the bands and the single-molecule and few-molecule crystals; the molecular axes are normal to the kebabs (parallel to the fibril axis), with the molecules folded. Extended sintering time with the presence of sufficient material resulted in the kebabs becoming (or growing as) bands (Figs. 48a, 50c, 51a).

**Acknowledgements** The recent research described here was funded, in part, by W.L. Gore and Assocs., Inc. and the NSF Polymer Program through grants NSF-DMR 96-16255 and 02-34678. A portion of the research was carried out in the Center for Microanalysis of Materials, University of Illinois, which is partially supported by the US Department of Energy under grant DEFG02-91-ER45439.

# References

1. Bunn CW, Cobbold AJ, Palmer RP (1958) The fine structure of polytetrafluoroethylene. J Polym Sci 28:365
2. Speerschneider CJ, Li CH (1962) Some observations on the structure of polytetrafluoroethylene. J Appl Phys 33:1871
3. Speerschneider CJ, Li CH (1963) A correlation of mechanical properties and microstructure of polytetrafluoroethylene at various temperatures. J Appl Phys 34:3004
4. Geil PH (1963) In: Polymer single crystals. Interscience-Wiley, New York, chap IV. Published internally at DuPont in 1961
5. Geil PH, Anderson F, Wunderlich B, Arakawa T (1964) Morphology of polyethylene crystallized from the melt under pressure. Polym Sci A 2:3707
6. Melillio L, Wunderlich B (1972) Morphology of polytetrafluoroethylene extended chain crystals VIII. Kolloid Z Z Polym 250:417
7. Bassett DC (1981) Principles of polymer morphology. Cambridge University Press, Cambridge, p 167
8. Keller A (1957) A note on single crystals in polymers: evidence for a folded chain conformation. Philos Mag 2:1171
9. Symons NKJ (1963) Growth of single crystals of polytetrafluoroehylene from the melt. J Polym Sci 1:2843
10. Wunderlich B Melillio L (1968) Morphology and growth of extended-chain crystals of polyethylene. Makromol Chem 118:250
11. Butenuth G (1958) Zur struktur von polytetrafluoroäthylen. Verhandlber Kolloid-Ges 18:168
12. Whalen JW Wade WH, Porter JJ (1967) The surface area and particle structure of Teflon 6. J Colloid Interface Sci 24:379
13. Rahl FJ Evanco MA, Fredericks RJ, Reimschuessel AC (1972) Studies of the morphology of emulsion-grade polytetrafluoroethylene. J Polym Sci Part A2 10:1337
14. Seguchi T, Suwa T, Tamura N, Takehisa M (1974) Morphology of polytetrafluoroethylene prepared by radiation-induced emulsion polymerization. J Polym Sci Polym Phys Ed 12:2567
15. Grimaud E, Sanlaville J, Troussier M (1958) Dispersions de polytetrafluoroethylene. J Polym Sci 31:525

16. Herschleb JH (1960) The structure of polytetrafluoroethylene latex paticles. Paper presented at the Electron Microscopy Society of America meeting, Milwaukee (August)
17. Chanzy H, Smith P, Revol JF (1986) High-resolution electron microscopy of virgin poly(tetrafluoroethylene). J Polym Sci Polym Lett Ed 24:557
18. Chanzy H, Folda T, Smith P, Gardner K, Revol J (1986) Lattice imaging in polytetrafluoroethylene single crystals. J Mater Sci Let 5:1045
19. Folda T, Hoffmann H, Chanzy H, Smith P (1988) Liquid crystalline suspensions of poly(tetrafluoroethylene). Nature 333:55
20. Clark ES (1999) The molecular conformation of polytetrafluoroethylene: forms II and IV. Polymer 40:4659
21. Siegle JC, Muus LT, Lin T-P, Larsen HA (1964) The molecular structure of perfluorocarbon polymers II. Pyrolysis of polytetrafluoroethylene. J Polym Sci Part A 2:391
22. Shulman GP (1965) Thermal degradation of polymers I. Mass spectrometric thermal analysis. Polym Lett 3:911
23. Yuan B-L, Rybnikar F, Saha P, Geil PH (1999) Phase I and II crystals of poly(p-oxybenzoate). II. Effects of solution polymerization conditions on morphology and crystal structure. J Polym Sci Polym Phys Ed 37:3532
24. Clark ES, Muus LT (1962) Partial disordering and crystal transition in polytetrafluoroethylene. Z Kristallogr 117P:119
25. Phillips DO (1988) Dependence of polyethylene melt viscosity on morphology and time. MS thesis. University of Illinois at Urbana-Champaign
26. Kent SL, Geil PH (1992) Folded chain crystallization of random terpolymer liquid crystal polymers from the nematic state. J Polym Sci Polym Phys Ed 30:1489
27. Kent S, Rybnikar F, Geil PH Carter JD (1994) Morphology of a thermotropic random terpolymer liquid crystal polymer crystallized in the bulk: compression mouldings, extrudates and fibres. Polymer 35:1869
28. Hussein IA, Williams MC (2004) Melt flow indexer evidence of high-temperature transitions in molten high-density polyethylene. J Appl Polym Sci 91:1309
29. Geil PH, (2001) Some "overlooked problems" in polymer crystallization. Polymer 41:8983
30. (a) Wunderlich B, Davison T (1969) Extended-chain crystals. I. General crystallization conditions and review of pressure crystallization of polyethylene. Polym Sci A2 7:2043; (b) Bassett DC, Block S, Piermarini GJ (1974) A high-pressure phase of polyethylene and chain-extended growth. J Appl Phys 45:4146; (c) Hikosaka M (1990) Unified theory of nucleation of folded-chain crystals and extended-chain crystals of linear-chain polymers: II. Origin of fcc and ecc. Polymer 31:458
31. Matsushige K, Enoshita R, Die T, Yamauchi N Taki, S Takemura, T (1977) Fine structure of the III-I transition and molecular motion in polytetrafluoroethylene. Jpn J Appl Phys 16:681
32. Weeks JJ, Eby RK, Clark ES (1981) Crystal structure of the low temperature phase (II) of polytetrafluoroethylene. Polymer 22:1496, and references therein
33. de Rosa C, Guerra G, Petraccone V, Centore R, Corradini P (1988) Temperature dependence of the intramolecular disorder in the high-temperature phase of poly(tetrafluoroethylene) (phase I). Macromolecules 21:1174
34. Clark ES, Weeks JJ, Eby RK (1980) Diffraction from non-periodic structures, the molecular conformation of polytetrafluoroethylne (phase II). In: French AD, Gardner KH (eds) Fiber diffraction ACS symposium series 141. American Chemical Society, Washington, DC, p 183
35. Gore RW (1976) Process for producing porous products. US Patent 3,953,566

36. Wang X, Zho J-J, Li L (2004) Twisting and bending of polyethylene crystals. Paper presented at PP' 2004, Dali (June)
37. Burmester A, Geil PH (1972) Small angle diffraction from crystalline polymers. In: Pae RD, Morrow DR, Chen Y (eds) Advances in polymer science and engineering. Plenum, New York, pp 42–100
38. Wittmann JC, Lotz B (1990) Epitaxial crystallization of polymers on organic and polymeric substrates. Prog Polym Sci 15:909

Adv Polym Sci (2005) 180: 161–194
DOI 10.1007/b107237
© Springer-Verlag Berlin Heidelberg 2005
Published online: 29 June 2005

# Morphological Implications of the Interphase Bridging Crystalline and Amorphous Regions in Semi-Crystalline Polymers

Sanjay Rastogi[1,2] (✉) · Ann E. Terry[3]

[1]Department of Chemical Engineering/Dutch Polymer Institute,
Eindhoven University of Technology, Den Dolech 2, P.O. Box 513, 5600MB Eindhoven,
The Netherlands
*s.rastogi@tue.nl*

[2]Max Planck Institute for Polymer Science, Ackermannweg 10, 55128 Mainz, Germany
*s.rastogi@tue.nl*

[3]Department of Chemical Engineering/Dutch Polymer Institute,
Eindhoven University of Technology, Den Dolech 2, P.O. Box 513, 5600MB Eindhoven,
The Netherlands
*a.e.terry@tue.nl*

**Abstract** In semi-crystalline polymers a range of morphologies can be obtained in which a chain may traverse the amorphous region between the crystals or fold back into the crystals leading to adjacent or nonadjacent reentry, depending on the molecular architecture and crystallization conditions. This causes topological variations on the crystal surface and the occurrence of an interphase between the crystalline and amorphous domains, thus affecting the mechanical properties. In this chapter, we will discuss how the morphology within the interphase plays a prominent role in drawability, lamellar thickening and melting of thus crystallized samples. Normally, for linear polymers it is anticipated that extended chain crystals are thermodynamically most favorable, and ultimately, taking the example of linear polyethylene, it has been shown that such chains would form extended chain crystals. However, this condition will not be realized in a range of polymers upon crystallization from the melt, such as those which do not show lamellar thickening or in branched polymers where the side branches cannot be incorporated within the crystal and hence fully extended chains are not possible. From a series of experiments, it is shown that with sufficient time and chain mobility, although extended chain crystals are not achievable, the chains still disentangle and a thermodynamically stable morphology is formed with a disentangled crystallizable interphase.

## Abbreviations

| | |
|---|---|
| 2D | two dimensional |
| DSC | differential scanning calorimetry |
| ESLD | ethylene sequence length distribution |
| CRF | crystalline rigid fraction |
| F2 | once-folded |
| FWHM | full width at half maximum |
| IR | infrared |
| LAM | Raman longitudinal acoustic modes |
| LLDPE | linear low-density polyethylene |
| MAF | mobile amorphous fraction |
| Mw | molecular weight |
| NIF | non-integer fold |
| nm | nanometer |
| NMR | nuclear magnetic resonance |
| PET | poly(ethylene terephthalate) |
| PEN | poly(ethylene naphthalate) |
| PBT | poly(butylene terephthalate) |
| $Q_0$ | equilibrium triple point |
| RAF | rigid amorphous fraction |
| SAXS | small-angle X-ray scattering |
| $T_c$ | crystallization temperature |
| $T_g$ | glass transition temperature |
| $T_m$ | melting temperature |
| UHMW-PE | ultrahigh molecular weight polyethylene |
| WAXD | wide-angle X-ray diffraction |

# 1
# Introduction

The semi-crystalline structures often formed by crystallizable polymers are known to consist of thin crystalline lamellae separated by amorphous regions [1-3]. For crystallization from the melt, where the conditions are far from equilibrium, the polymer chains must achieve a regular conformation from the highly entangled melt and align parallel to each other to form thin plate-like lamellae, the entanglements being confined to the amorphous regions. It is still unclear whether the polymer chains actually do disentangle or merely that during crystallization, entanglements are pushed to the surface. However, independent of the mechanism involved, the molecular structure of the amorphous region strongly depends not just on the chemical nature and inherent shape of the polymer, but also on the crystallization conditions either in the quiescent state or obtained during flow. Experimental efforts to decouple the mechanical properties of polymers from the crystalline and amorphous fractions have not been particularly successful because of this dependence of the molecular organization in the amorphous region upon the crystallization conditions. Moreover, a third structural component, an interphase of intermediate order, could exist between the amorphous and crystalline phases, as has been proposed both theoretically and experimentally, which means that a sharp demarcation line between the amorphous and crystalline phases is unlikely. This added complication is necessary if one considers that the chains emerge at the crystalline surfaces with a high degree of molecular alignment. These chains must either fold back into the crystallite, by adjacent reentry, or must reside in the amorphous matrix. In the case of crystallization from the melt, where the crystallization conditions are often very far from equilibrium, extensive and perfect folding will strongly depend on molecular weight and molecular architecture – and in most cases will be highly improbable. Assuming the crystallites are of an infinite extent in the basal plane, then at small distances away from each crystallite surface most of the chains present will have originated from the crystallite. The average chain orientation here will not be random as in the bulk amorphous matrix but will be distributed around the normal to the crystallite surface: this is the proposed semi-ordered interfacial component, the degree of ordering strongly depending upon the crystallization conditions. In several cases this may also give rise to a crystallographic register and further influence the physical and mechanical properties of the polymer.

A comprehensive review of the subject, supporting the existence of this third component in the structure of unoriented, semi-crystalline polymers, has been compiled by Mandelkern [4], and has been further supported by WAXD studies by Windle [5]. However, this still remains a topic of controversy. Terms used to describe the suggested third structural component have included 'semi-ordered', 'intermediate', 'rigid amorphous', 'interfacial', 'interzonal', 'interphase' and 'transitional zone', and give quite a clear picture of what is envisaged (for this review, we will use the term 'interphase'). Figure 1, reproduced from the work of Rutledge and coworkers [6], illustrates the molecular picture of the interphase, in which three types of chain populations exist: *bridges* that join two crystal lamellae, *loops* that have their entry and exit points on the same crystal lamellae, and *tails* that terminate in the amorphous phase.

At this stage it is essential to recapitulate the existing knowledge on chain tilting and its influence on the interphase. It is well established that polyethylene shows a crystalline lamellar structure and that there is chain folding in the interphase region, one of the important observed features being that the polymer chain stems are not generally orthogonal to the basal plane of the lamellae. Such basal planes are identified by a chain tilt angle, defined as the angle between the lamellar normal and the *c*-axis of polymer chain stems. The in situ process of chain tilting during heating has been shown previously and it is known that the initial chain tilt angle is strongly dependent on crystallization conditions. For example, Bassett and Hodge [7] studied polyethylene spherulites using electron microscopy and found a regular texture showing well-defined lamellae having crystal stems inclined at an angle ranging from 19° to 41° to the lamellar normal, with 34° (corresponding to the {201} facet) being the most common. Khoury [8] observed the presence of a predominant chain tilt angle of approximately 34° in polyethylene spherulites grown from the melt at high undercoolings. Under special conditions chain tilt angles higher than 45° could also be observed.

Chain tilting is known to exert a strong influence on the structure and hence the properties of the interphase. Frank suggested that mutual exclusion imposes steric constraints on the structure of the interphase between crystalline and amorphous regions that is only relieved by the presence of chain ends, folding back into the crystal and chain tilt [9]. Yoon and Flory [10] and Kumar and Yoon [11] also considered that the flux of chains into the amorphous region would be reduced by the tilting of the chains and the effect of this on chain folding. Mandelkern [12] also suggested that chain tilt angle is an important factor in any detailed description of the mechanism of crystallization and that it also influences the interfacial free energy associated with the basal plane. Bassett et al. [13] and Keith and Padden [14] discussed the role of tilted growth of polymer crystals in allowing better pack-

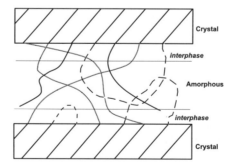

**Fig. 1** Schematic diagram of the interphase between crystalline domains, showing the examples of loops (*dashed*), bridges (*dotted*), and tails (*solid*), reproduced from Gautam et al. [6]

ing and enhanced space for the accommodation of disordered conformations and relatively bulky loops within interfacial layers. Recently, Toda et al. [15] have conclusively shown that the direction of chain tilting in polyethylene single crystals is different for each growth sector, which results in a selected handedness to the spiral terraces during crystal growth.

Using off-lattice Monte Carlo simulations, Rutledge and coworkers [6, 16] have reported a recent, detailed study on the tilted-chain interphase. Their simulations reveal the thermodynamic properties for a metastable interphase with different degrees of chain tilt. The interphase was considered to contain three types of chain populations, as shown in Fig. 1. Although the density decreases as one moves from the crystalline to the amorphous phase, an increase or peak in the density is observed around 0.6 nm away from the crystal surface. In the same manner, the order parameter, which starts at unity in the crystal, drops smoothly to zero as one moves into the amorphous except for a marked drop at a distance of $\sim 0.6$ nm from the crystal face. The results from density and orientational order were further supported by the occurrence of a transverse structure parallel to the crystal surface. This is explained by a higher number of loops in the interphase, while tails and bridges contribute more to the amorphous region, with a well-defined fold surface near 0.6 nm, at which a large number of chains fold back into the crystal. The transverse structure observed indicates a number of chains running parallel to the crystal surface. Since a large areal density (or flux) of chains leave the crystal surface, the interphase generates a fold surface in order to reduce the flux to a level sufficient to accommodate disordering. Crystallographic planes could be assigned to the fold surface.

The findings described above form a basis for our current work that examines the influence of the interphase on the structure and ultimately on the properties of linear polymers. In this chapter we will first show that, depending on the crystallization conditions, the amount of loops or entanglements in the interphase, and thus the deformation behavior of polymers, can be varied. We will initially consider the example of ultrahigh molecular weight polyethylene (UHMW-PE) and the role of entanglements upon its drawability.

# 2
# Control of entanglement density upon crystallization

As we have outlined above, it is important to consider the role of entanglements within the interphase. Entanglements as such are ill-defined topological constraints, which are usually visualized in textbooks as four strands leading away from a mutual contact, see Fig. 2.

Entanglements can be removed effectively by dissolution in a good solvent and by slow (isothermal) crystallization from the melt or from solution.

**Fig. 2** Simple picture of an entanglement

Slow crystallization promotes disentangling since the process of 'reeling-in' of chains from the entangled melt onto the crystal surface is promoted and this in turn enhances the maximum drawability, despite the fact that this increases the crystallinity. However, there is a critical lower limit for the number of entanglements between crystals in order to achieve high drawability and in the extreme limit of very slow isothermal crystallization, linear standard polyethylenes (not the UHMW-PE grades) can lose their drawability completely and become brittle materials.

## 2.1
### Crystallization via dilute solution

The way to remove entanglements, viz. the manner in which topological constraints limit the drawability, is seemingly well understood and crystallization from semi-dilute solution is an effective and simple route to make disentangled precursors for subsequent drawing into fibers and tapes [17, 18]. A simple 2D model visualizing the entanglement density is shown in Fig. 3. Here $\phi$ is the polymer concentration in solution and $\phi^*$ is the critical overlap concentration for polymer chains.

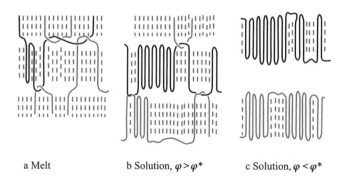

a Melt                          b Solution, $\varphi > \varphi^*$                          c Solution, $\varphi < \varphi^*$

**Fig. 3** A simple 2D model envisaging how the entanglement density varies upon crystallization at decreasing polymer concentration, $\phi$. $\phi^*$ is the critical overlap concentration for polymer chains

Upon crystallization from solution, at $\phi > \phi^*$, the molecules usually fold along a specific plane. The stems of a test chain (filled dots) are shown in Fig. 4 without indicating the folds. For the sake of simplicity, we assume that adjacent reentry occurs during crystallization and that the chain is located within one crystal plane, thus forming a well-organized disentangled crystal surface. The low-shear moduli of polyethylene crystals means that chain mobility in the direction perpendicular to the chain and along the $\{110\}$ plane is rather high. In contrast, upon crystallization from the melt the stems of the test chains are crystallized with a lower degree of order and a highly entangled interphase is formed. In this case, shearing (slip) is more difficult since the chains cannot cross each other during deformation.

Independent of the crystallization route taken, the role of entanglements upon the mechanical properties will be the same, i.e. acting as physical crosslinks on the time-scale of drawing experiments of semi-crystalline polymers. This can be demonstrated by comparing the deformation of solution-cast UHMW-PE films to melt crystallized films. From the stress–strain curve shown in Fig. 5, it is evident that above the $\alpha$-relaxation temperature and below the melting temperature, solution-cast films do not show strain hardening and can be drawn to more than 40 times the original length. In contrast, melt-crystallized films show strain hardening and could be drawn only by six to seven times. This difference in the strain-hardening behavior can be attributed directly to the difference in entanglements present as described above.

The schematic representation of stems within the crystals is, of course, an oversimplification to explain the drawing behavior of UHMW-PE films. In practice, superfolding and the crossing of stems belonging to the same chain will occur. The presented model, however, serves to demonstrate that adjacent reentry and the locality of molecules within a crystal will cause a structured interphase to form. This will facilitate the process of ultra-drawing, comprising the breaking of lamellar crystals via shearing, tilting and subsequent

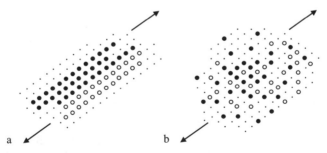

**Fig. 4** Simplistic representation of a chain fold in a polymer crystal. The dots represent stems of molecules folding along the $\{110\}$ plane, viewed along the $c$-direction. No folds are drawn for the sake of simplicity

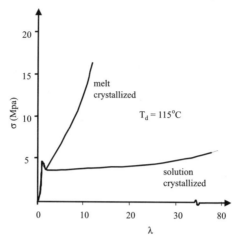

**Fig. 5** Stress–strain curves for solution-cast UHMW-PE films compared to melt-crystallized films. $\lambda$ refers to the draw ratio. Films were drawn at a drawing temperature of $T_d = 115\,°C$

unfolding of clusters. The subsequent instantaneous loss in drawability upon melting and recrystallization is due to the rearrangement and intermixing of stems involving only local chain motions rather than movement of the complete chains as proposed for self-diffusion in polymer melts.

## 2.2
## Exploitation of the hexagonal phase in polyethylene

Another way to disentangle linear polyethylenes, and thus control the interphase without using a solvent, is to anneal the polymer in the hexagonal phase. Bassett has discussed the role of the hexagonal phase in the crystallization of polyethylene extensively in an earlier chapter in this book. Briefly, polyethylene exhibits a number of different crystal structures, with the hexagonal phase being observed in linear polyethylenes at elevated pressure/temperature in isotropic samples or at ambient pressure in oriented samples. For this reason, we have to distinguish between these two situations, namely isotropic and oriented polyethylene. However, we will focus only on isotropic polyethylene and will refer readers to reference [18, 19] for an overview of oriented polyethylene.

For isotropic polyethylene, the hexagonal phase is usually observed at elevated pressure and temperature, in fact, above the triple point $Q$ located at 3.4 kbar and 220 °C according to the pioneering work of Bassett et al. [20] and Wunderlich et al. [21]. Later, more detailed studies involving in situ light microscopy and X-ray studies showed that the equilibrium point, $Q_0$, is located at even higher temperatures and pressures, 250 °C and 5.3 kbar, re-

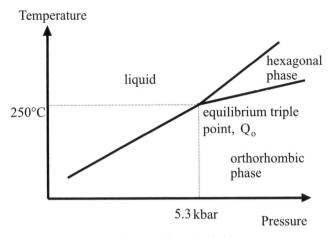

**Fig. 6** Pressure–temperature phase diagram for polyethylene

spectively, see Fig. 6 [22–24]. The hexagonal phase is a so called mobile phase with a high degree of chain mobility along the $c$-axis. Annealing in the hexagonal phase promotes lamellar thickening with the formation of extended chain crystals via disentangling of chains and consequently aids drawability. For UHMW-PE (according to ASTM definitions $M > 3.10^6$ D) it is therefore possible to disentangle the chains even in the solid state by exploiting the hexagonal phase. The resulting organized molecular structure in the amorphous phase enhances the subsequent drawing operation, as is shown by Ward et al. [25].

## 2.3
## Via synthesis

An additional elegant route to obtain disentangled UHMW-PE, and thus to control the interphase, is by direct polymerization in the reactor. In order to make UHMW-PE, a relatively low polymerization temperature is needed and a situation is easily encountered where the polymerization temperature is lower than the crystallization temperature of UHMW-PE in the surrounding medium in which the catalyst is suspended. In this situation, the growing chains on the catalyst surface tend to crystallize during the polymerization process. These UHMW-PE reactor powders, often referred to as nascent or virgin UHMW-PE, can be remarkably ductile. It was shown by Smith et al. [26] that reactor powders, in the same manner as solution-cast UHMW-PE, could be drawn easily into high-modulus structures.

Again the proposed entanglement model can be invoked to explain the ductility of compacted UHMW-PE reactor powders. The growing chains on the crystal surface can crystallize independently of each other and consequently disentangled UHMW-PE is obtained as a direct result of polymerization. Depending on the polymerization conditions, the crystal size of the nascent morphology can also be controlled. As with the melting-point dependence of polyethylene on the crystal thickness (fold length), the triple point, $Q$, in the pressure–temperature phase diagram of polyethylene is also dependent on the crystal dimensions. In particular, for nascent UHMW-PE reactor powders that consist of crystallites with very small dimensions, it was shown that a metastable hexagonal phase could be observed at pressures and temperatures as low as 1 kbar and 200 °C, respectively [27, 28] and upon annealing a transformation into the thermodynamically stable orthorhombic phase occurs. The observation that a thermodynamically stable crystal structure is reached via a metastable state of matter is not unique for polyethylene, nor for polymers; indeed it has been invoked as early as in 1897 by Ostwald, commonly expressed as Ostwald's stage rule [29]. For polyethylene, it has been shown that crystals at elevated pressures initially grow in the hexagonal phase and after a certain time, or once a certain crystal size has been attained, these hexagonal crystals are transformed into thermodynamically stable orthorhombic crystals.

Figure 7 shows electron micrographs of two different virgin UHMW-PEs. Electron micrographs clearly show that the lamellae in the laboratory synthesized sample, A, thicken substantially for the same annealing conditions, for example pressure, temperature and time, compared to a commercially synthesized grade, sample B. There is a marked difference in their drawability; the laboratory grade can be drawn in the solid state below the melting point, whereas the commercial one cannot. Details of molecular weight, molar-mass

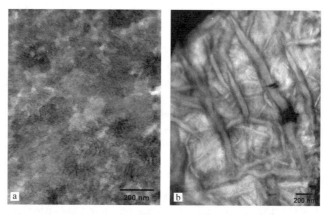

**Fig. 7** Electron micrographs of two different virgin UHMW-PE samples, (**a**) commercially synthesized Z-N (grade B) and (**b**) laboratory-synthesized Z-N (grade A). The samples were annealed in the hexagonal phase, at 1500 bars, 190 °C for 30 minutes. Under these conditions the samples were in the thermodynamically metastable hexagonal phase. (For details see references [27, 28])

**Table 1** Molecular weight, molecular weight distribution, polymerization temperature

|  | $M_w$ [g mol$^{-1}$] | $M_w/M_n$ | $T_{synthesis}$ |
| --- | --- | --- | --- |
| Controlled synthesis, Z-N (Grade A) | $3.6 \times 10^4$ | 5.6 | 50 °C |
| Commercial, Z-N (Grade B) | $4.54 \times 10^4$ | 10.0 | 80 °C |

distribution, polymerization temperature are given in Table 1. However, the difference in drawability arises due to the reduction in the number of entanglements on the fold surface of the laboratory grade sample compared to those present in the commercial grade. This highlights once again the influence of entanglements and the interphase on these materials and their effect on the molecular-chain mobility even during lamellar thickening. The influence of entanglements on lamellar thickening has been probed further by solid-state NMR.

# 3
# Influence of the interphase on molecular mobility

Recently, with the help of solid-state NMR, Uehara et al. [30] investigated the role of entanglements present in the interphase on molecular-chain mobility along the c-axis of the crystal lattice of UHMW-PE. They used solution- and melt-crystallized films and nascent powders, observing regular lamellar stacking in the solution-crystallized films, whereas the nascent powders and melt-crystallized samples showed conventional non-stacked lamellar morphology. By $^1$H pulse NMR measurements, they defined three different relaxation processes occurring during heating. In process 1 (heating from room temperature), an activation of molecular motion at the boundary between crystal and amorphous regions takes place. During process 2 (above 60–90 °C, i.e. the α-relaxation temperature), the crystallinity increases with an acceleration of the entire molecular motion caused by sliding of molecular chains in the crystalline region. Raising the temperature further above 130 °C (process 3) leads to the start of sample melting. For solution-grown crystals and nascent powder samples, the crystalline relaxation exhibited all three processes as well as constraint of the amorphous chains; however, the transition between processes 1 and 2 occurred at a higher temperature for the nascent morphology. In contrast, melt-grown crystals did not show process 1 and directly led to process 2 upon heating. This therefore suggests that the accelerated molecular motion in the crystal–amorphous boundaries occurs following lamellar thickening via a solid-state reorganization, i.e. without melting, for the highly crystalline nascent and solution-crystallized samples having fewer entangled chains in the amorphous phase. For melt-crystallized samples, the higher density of entanglements trapped on the lamellar surface precludes such a boundary relaxation; thus, lamellar thickening may not occur. These results imply that the trapped entanglements also play an important role in lamellar rearrangement during annealing such as might occur during welding of semi-crystalline polymers.

## 4
## From the interphase to the interface:
## the welding of semi-crystalline polymers

By considering the influence of the interphase upon annealing, it is possible to shed a little light on the welding behavior of semi-crystalline polymers, which has received much less attention than that of amorphous polymers. Because of the ill-defined morphology of the interphase the welding characteristics of semi-crystalline polymers are quite different from amorphous polymers and are far from well understood.

In the case of solution-crystallized UHMW-PE, it is possible to make solution-cast films in which the lamellar crystals are regularly stacked,

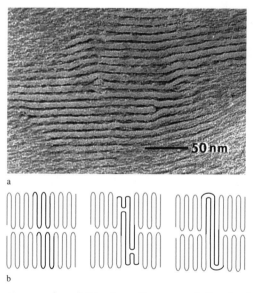

**Fig. 8** (**a**) Electron micrograph and (**b**) schematic representation for the regular lamellar structures formed within solution-cast films of UHMW-PE

see Fig. 8. Upon heating these (dry) solution-cast films above approximately 110 °C, it has been observed, by in situ synchrotron measurements and time-resolved longitudinal acoustic-mode raman spectroscopy [31], that the lamellar thickness increases finally to twice its initial value, from 12.5 nm to 25 nm, with the loss of the well-stacked lamellar arrangement after the doubling process. Figure 8b shows the model for the chain rearrangement during heating. However, the drawability of the annealed solution-crystallized films was maintained after lamellar doubling since no stem intermixing, or entangling, occurs as discussed previously in the stem-rearrangement model, Fig. 4.

A well-defined amount of co-crystallization is possible across the interface of two adjacent crystals by annealing two stacked, completely wetted, solution-cast films of UHMW-PE [32]. It was found that doubling of the lamellae across the interface enhances the peel energy to such a level that the films could not be separated anymore. By contrast, 'pre-annealing' one side of the film prohibited co-crystallization across the interface and these films could still be separated easily. It was therefore concluded that a limited amount of chain diffusion across the interface occurs during doubling of the lamellae, as facilitated by the well-defined structure of the interphase due to the adjacent reentry that occurs upon crystallization from solution.

Thus far we have considered the influence of the interphase in relation to the mechanical properties of semi-crystalline linear polymers, in particular for the case of UHMW-PE, i.e. its role in the drawability or in the welding of such materials. With the recent advent of synthesis routes to produce highly controlled model systems, the investigation of the interphase can be extended further.

# 5
## Influence of chain folding on the unit cell

Normally, for linear polymers it is anticipated that extended chain crystals are thermodynamically the most favorable, and ultimately given an example of linear polyethylene, it is shown that chains within the folded chain crystals tend to move along the $c$-axis via chain sliding diffusion to attain a thermodynamically stable morphology. However, the possibility of chain diffusion within the crystal lattice cannot be realized normally when the molecular structure changes from linear to branched, in particular, where the side branches cannot be incorporated within the crystal. Below, we have summarized our recent studies on such branched polymers and will demonstrate a thermodynamically stable morphology that these polymers ultimately tend to achieve. To reach the objective, it is essential to recapitulate first our findings on model systems, for example, linear and branched ultra-long alkanes, which are essential for an understanding of the experimental observations of branched polyethylenes, for example, ethylene–octene copolymers.

It has been known since the early 1970s that the thickness of polyethylene lamellae can have some influence on the lattice parameters of the unit cell, with a tendency towards higher density at larger crystal thickness. Further studies of this effect are hindered in the case of higher molecular weight polymers by the polydispersity of the materials. Polymers are mixtures of chains of different lengths and the lack of purity is likely to lead to a higher level of defects than would occur in a pure system. This gives a very strong dependence of structure on the age of the crystal and the way in which it was crystallized. During the past two decades ultra-long monodisperse alkanes, with chain lengths up to 390 carbons, have become available. These materials, which are model substances for low molecular weight polymers, have provided many new insights into polymer crystallization in general and polyethylene crystallization in particular.

## 5.1
### Monodisperse ultra-long linear alkanes

Ungar and Zeng [33] have comprehensively summarized the research on strictly monodisperse materials from their first synthesis in 1985 until 2001. From the earliest studies it became apparent that, due to the monodispersity of the materials, the thickness of the lamellar crystals formed is always an integer fraction of the extended chain length (allowing for any chain tilt), such that the polymers always crystallize in the extended chain form or fold exactly in half (once-folded), or in three (twice-folded), etc. This behavior means that, when the alkanes are crystallized at a particular temperature, the entire lamellar population has very closely the same thickness and stability. The use of such an ultra-pure system to study the impact of thickness on lattice parameters removes many of the problems inherent to polymers, whilst maintaining the most important characteristic of chain length.

What follows is a brief overview of wide-angle X-ray diffraction study using the high-resolution time-resolved capabilities of the European Synchrotron Radiation Facility to investigate the effect of chain folding and unfolding on lattice parameters in a series of five ultra-long alkanes of different chain lengths: $C_{102}H_{206}$, $C_{122}H_{246}$, $C_{198}H_{398}$, $C_{246}H_{494}$ and $C_{294}H_{590}$ [34]. Both $C_{102}H_{206}$ and $C_{122}H_{246}$ can only be obtained in the extended chain form under typical dilute solution crystallization conditions; all the other alkanes undergo chain folding under some crystallization conditions. Table 2 shows the samples used and the crystal thicknesses obtained, assuming six carbons per fold (these thicknesses agree well with the thicknesses measured on similarly prepared crystals using Raman LAM and small-angle X-ray scattering). The effect of heating at $2\,°C/min$ on the lattice parameters of these alkanes was examined.

**Table 2** The ideal crystal thickness for differently folded forms of the alkane samples studied, assuming six carbons per fold

| Alkane sample | Crystal thickness, nm |
|---|---|
| $C_{102}H_{206}$ extended | 13.2 |
| $C_{122}H_{246}$ extended | 15.7 |
| $C_{198}H_{398}$ extended | 25.4 |
| $C_{198}H_{398}$ once-folded | 12.0 |
| $C_{246}H_{494}$ extended | 31.52 |
| $C_{246}H_{494}$ once-folded | 15.1 |
| $C_{294}H_{590}$ once-folded | 18.2 |
| $C_{294}H_{590}$ twice-folded | 11.2 |

Examining the lattice parameters in detail by fitting the (110) and (200) peaks provides three separate but interrelated sets of information. Firstly, as expected, the lattice expands on heating due to the increased thermal motion of the chains. In the case of the extended chain crystals, which are already close to equilibrium (at least with respect to size), this is all that happens; see Fig. 9. Constant thermal expansion of the a and b lattice parameters can be seen, the coefficients of thermal expansion derived from this data are $4 \times 10^{-3}$ Å/°C for the $a$-axis, and $-3 \times 10^{-4}$ Å/°C for the $b$-axis, these basically remaining constant for each of the chain lengths examined and, as

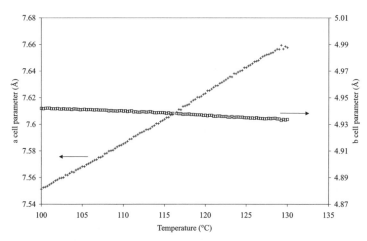

**Fig. 9** The variation in the a(+) and b(□) cell parameters for $C_{246}H_{494}$ extended chain crystals. The left-hand axis refers to the a cell parameter and the right-hand axis to the b cell parameter. Just prior to melting the cell parameters drop, probably due to an increase in the error as the peak intensity rapidly drops to zero, reducing the accuracy of the peak-fitting routine

expected, the final melting point is at a higher temperature for the longer chain molecules. Secondly, if the crystals consist of folded chains, the thickening of these crystals that occurs above 120 °C is accompanied by a contraction in the crystal lattice. Figure 10 shows the contraction in lattice parameters for the once-folded form of $C_{246}H_{494}$, as shown previously for the extended chain form in Fig. 9. The contraction during thickening is very striking, and even more so in the case of $C_{294}H_{590}$ where it occurs on the transition both from twice-folded to once-folded crystals, and at the transition from once-folded to extended chain crystals, Fig. 11. This is the first time that such a transi-

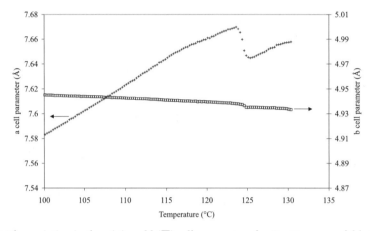

**Fig. 10** The variation in the a(+) and b(□) cell parameters for $C_{246}H_{494}$ once-folded chain crystal. The left-hand axis refers to the a cell parameter and the right hand axis to the b cell parameter

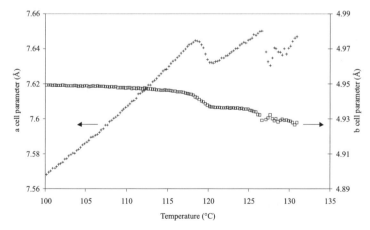

**Fig. 11** The variation in the a(+) and b(□) cell parameters for $C_{294}H_{590}$ twice-folded crystals. The left-hand axis refers to the a cell parameter and the right hand axis to the b cell parameter

tion has been followed in real time using WAXD, although the transformation to a less-folded form had been measured previously by DSC. The contraction of the lattice that accompanies unfolding is superimposed on top of the thermal expansion. Finally, prior to this contraction, there is an increase in the full width at half maximum (FWHM) of each of the crystal Bragg peaks over a temperature range of several degrees, which can be associated either with an increase in local strain in the crystals, or a reduction in crystal size in the lateral direction.

The use of different alkanes with different crystal thicknesses and numbers of folds within the crystal enables us to assert with confidence that the contraction in the lattice is due to the thickening process. The fact that the alkanes, due to their strict monodispersity, form crystals with only a few closely defined thicknesses enables this contraction to be seen much more clearly than it is in polyethylene. In polyethylene a range of different crystal thicknesses may be present, and the thickening process generally occurs over a wider range of temperatures, smearing out any step-like effect that may exist.

It may be concluded that a rapid contraction of the crystalline lattice with morphological changes at the surface of the crystal, and hence the interphase, occurs as ultra-long, monodisperse alkanes undergo a transition between different integer-folded forms on heating and that this contraction takes place in all the materials studied, despite apparently different routes being taken between achieving the pre- and post-transition crystal forms. The different lattice parameters can be associated with a particular crystal thickness and fold surface density and the observed contraction of the lattice is remarkably clear evidence of the effect of surfaces on the polymer crystal structure.

The extent to which the thickening process occurs by solid-state reorganization of the parent crystals or by melting and recrystallization into thicker crystals is still a matter of discussion and probably depends on the polymer in question and the annealing temperature. Chain mobility within lamellae has been investigated by solid-state NMR [35]. Recent studies have also pointed to the possible role of mobile phases, such as the hexagonal phase in polyethylene, enabling the thickening process to occur. Computer simulations have also been used to address these issues, although this is still hampered by the cooperative nature of the process and therefore the requirement for a large simulation size. In the WAXD study of lamellar thickening by Terry et al. [34], the intensity of the diffraction peaks allows any changes in the degree of crystallinity to be monitored. It was found that, in the case of unfolding of $C_{198}H_{398}$ from once-folded to extended and the unfolding of $C_{294}H_{590}$ from twice-folded to once-folded, there was little change in the intensity during the transition. This implies that the transition must be a solid-state process, or that any melting is very localized. However, other transitions, for example the unfolding of $C_{294}H_{590}$ from once-folded to extended, showed that the material almost melts completely and then recrystallizes into the thickened form.

This difference in behavior must reflect the difference in energetic barrier in the different cases, perhaps due to differences in the degree of thickening required in the transition between the different forms, the proximity to the equilibrium melting point and also perhaps to differences in the initial lamellar thickness.

Thus far, we have only considered studies using simple, highly crystalline linear $n$-alkanes. Introducing short chain branches into the amorphous region of the crystals further influences the interphase and the corresponding lattice parameters. Two monodisperse branched $n$-alkanes were also synthesized by Brooke et al. [36], $C_{96}H_{193}CH(R)C_{94}H_{189}$ where R = $CH_3$ and $(CH_2)_3CH_3$, i.e. a methyl and a butyl branched alkane, respectively. The crystallization of polyethylene copolymers, for which these branched $n$-alkanes serve as an analogue, is highly complex [37]. Some of the key differences of the molecular organization of branched polymers compared to linear alkanes during crystallization are now described.

## 5.2
### Monodisperse ultra-long branched alkanes

When examining the crystallization behavior of branched alkanes, consideration should also be given to whether the branches due to their length will be excluded to the crystal surface, as we have mentioned earlier. It is generally accepted that methyl branches can be incorporated at interstitial sites, leading to distorted lamellae [38–40], whereas hexyl branches are definitely rejected from the crystalline core. Whether ethyl branches are included or not depends upon the cooling rate used for crystallization; at higher cooling rates, the possibility of incorporation into the crystal is increased. A further complexity is that the ethylene sequences, if sufficiently long, will be able to fold, however, this is unlikely to be a tight fold, rather a longer fold is envisaged in order to keep the comonomers out of the crystal. As the concentration of comonomers increases, so the lamellar thickness will decrease and overcrowding of the branches in the amorphous region will result [7, 41, 42]. This overcrowding is relieved either by the lamellae curving, the chains in the crystal tilting or even by limiting the lateral size of the crystals. Consequently, thin, imperfect crystals are formed under such conditions.

Detailed SAXS studies using synchrotron radiation on both monodisperse linear and branched $n$-alkanes have been performed by Ungar et al. [43–49] under isothermal and/or non-isothermal conditions at atmospheric pressure. The investigations unveil and explain the complex chain-folding process during crystallization. It was observed that the crystals formed in the early stages of crystallization have lamellae with the thickness of a non-integer fraction (NIF) of the extended form. This transient form soon transforms into an integer form, either via thickening at high annealing temperatures (to the extended form) or thinning at low annealing temperatures (to the once-folded form, F2). There is a lower entropic barrier for the random attachment of the chain in the NIF as compared to that for once-folded crystals where all end groups must be positioned at the crystal edge before deposition, and so the chain prefers the fast random attachment. There is evidence to suggest that this disordered layer contains long uncrystallized chain ends (cilia) from molecules which have 'half crystallized' and which are subsequently drawn into the crystals by a process of chain translation. This implies that there must be a high degree of chain mobility even within the crystalline lattice to allow such reorganization of the chains within the growth front, such mobility as was observed during the morphological changes that occur during crystal thickening of linear $n$-alkanes.

In the past few years, Ungar et al. have elucidated the structure and mechanism of the formation of the NIF form using SAXS (including electron density profiles), Raman longitudinal acoustic modes (LAM) spectroscopy and differential scanning calorimetry (DSC). A comparison of the crystallization mechanism of the linear and branched alkanes [47] indicated that in the sequence, melt $\rightarrow$ NIF $\rightarrow$ F2, the melt $\rightarrow$ NIF step is fast but the NIF $\rightarrow$ F2 step is slow in linear alkanes. The crystallization mechanism is understood in the following manner. In NIF lamella of a linear alkane, a half-crystallized molecule generally has two cilia. During crystallization, neither of the two cilia is long enough to make a complete adjacent reentry, provided no rearrangements in the already crystallized part occur. This retards the formation of F2 crystals. However, the reverse is true for the branched alkane. An isothermal crystallization study of the methyl branched alkane $C_{96}H_{193}CH(CH_3)C_{94}H_{189}$ revealed that the high rate of NIF $\rightarrow$ F2 transition is attributed to the fact that, here, the only successful deposition mode of the first stem (first half) of the molecule is the one which places the branch at a lamellar basal surface and the chain end at the opposite surface of the same lamella. The other half of the molecule (uncrystallized cilium) is then ideally suited to complete a second traverse of the crystal [47].

To gain a better insight into the crystallization mechanism and resulting crystal structures of the branched alkanes with respect to the branch length, the phase behavior of the butyl branched alkane $C_{96}H_{193}CH(C_4H_9)C_{94}H_{189}$ has been investigated at elevated pressures. A well-defined morphology is expected where the chains are adjacent reentrant, due the enhanced chain mobility along the $c$-axis at elevated pressures and temperatures. If we envisage the folding of a single molecule then the branch, which occurs exactly after 96 C-atoms and is followed by 94 C-atoms, will lie almost in the middle of the fold. Variations in the crystal structure have been followed in situ with WAXD while the morphological aspects have been investigated using in situ SAXS. The high-pressure measurements were done using a piston cylinder-type pressure cell similar to that designed by Hikosaka and Seto [50] capable of attaining a maximum pressure of 5.0 kbar and temperatures from room temperature to 300 °C.

At atmospheric pressure, the butyl branched alkane shows a similar packing to linear alkanes and linear polyethylene. Orthorhombic packing (and monoclinic due to shear in the sample) is maintained even though the branches will be excluded to the lamellar surface. For the as-synthesized sample, the lamellar spacing corresponds to chains that are once-folded and perpendicular to the basal plane, i.e. the chains are not tilted. Upon heating, some of the chains will begin to tilt with a chain tilt angle of approximately 35°, and upon cooling this tilted structure is retained. Non-integer folds were not observed. This suggests that during non-isothermal crystallization the chains tend to resort to the once-folded structure and one can assume that the butyl branch is excluded to the surface.

Initially, if one applies a pressure of 3.8 kbar to the sample, no change in behavior is seen compared to that observed in linear alkanes [51]. The intensity of the diffracted pattern will decrease but that is purely due to the thickness of the sample decreasing with increased pressure. As the sample is heated, the monoclinic phase will disappear first at approximately 160 °C, with a corresponding increase in the intensity of the orthorhombic peaks. The sample finally melts at approximately 258 °C.

Interestingly, the effects of the branches on the phase behavior and the significance of the interphase were clearly shown by cooling from the melt at elevated pressures, see Fig. 12. Upon crystallizing from the melt, crystal formation occurs directly in the orthorhombic phase. The orthorhombic (110) and (200) reflections gain intensity with increasing supercooling. A weak monoclinic reflection appears at approximately 148 °C. These reflections have been assigned considering the earlier work by Hay and Keller [52]. With subsequent cooling at $\sim$ 70 °C, a relatively broad and weak new reflection appears next to the monoclinic reflection. The appearance of the new reflection is followed by a sudden drop in the intensity of the orthorhombic reflection and a simultaneous increase in the intensity of the monoclinic reflection. The presence of the new reflection becomes more evident with further cooling. The (110) and (200) orthorhombic peaks show a sudden shift to higher angles implying a densification of the orthorhombic crystalline lattice with the appearance of the monoclinic reflection and the new reflection. Figure 12b shows that, after the appearance of the new reflection at $\sim$ 70 °C, a dramatic decrease along the $a$-axis of the orthorhombic unit cell occurs together with a sharp decrease along the $b$-axis. Densification of the orthorhombic unit cell ($\varrho = 1141$ kg/m$^3$ at 4.0 kbar, 25 °C) is evident from Fig. 12c.

Definite assignment of a phase to the new reflection is not straightforward as only one reflection is observed; for the present it has been termed a pseudo-hexagonal phase as its spacing is close to that of the hexagonal phase in polyethylene ($d = 4.16$ Å, c.f. $d_{100hex}$(polyethylene) = 4.2 Å). We already know that, due to their length, the butyl branches are rejected from the crystalline lattice and so are segregated to the fold surface of the crystal. It is suggested that, with pressure, this new phase appears when these

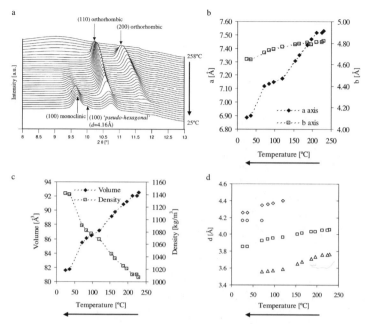

**Fig. 12** (**a**) A series of integrated WAXD patterns for the crystallization of butyl-branched alkane at elevated pressures of 4.0 kbar while cooling at a rate of 4 °C/min. The orthorhombic (110) and (200) reflections appear first, and gain intensity with increasing supercooling. A weak (100) monoclinic reflection appears at ~ 148 °C. Upon further cooling at ~ 70 °C, a sudden drop in the intensity of the (110) orthorhombic reflection is observed with the appearance of a new reflection and a simultaneous increase in the intensity of the (100) monoclinic reflection (**b**) Plot showing compression of the orthorhombic unit cell parameters during cooling at elevated pressures (**c**) A linear decrease and increase in the volume and density of the orthorhombic unit cell is respectively observed during cooling at ~ 4.0 kbar (**d**) Plot of Bragg d-values of various crystalline reflections versus temperature during cooling at elevated pressure of 4.0 kbar: (100) monoclinic (◇), (100) pseudo-hexagonal (○), (110) orthorhombic (□), (200) orthorhombic (△). The X-ray wavelength used for these experiments is 0.744 Å

butyl branches crystallize together. The separation of the butyl branches from each other is greater than that of the main chain within the crystal due to the fact that a fold occurs at the surface. The observed densification of the orthorhombic unit cell therefore acts to relieve the strain induced at the surface by the expanded nature of the crystallization of the branches. This is supported by the observations by SAXS, Fig. 13, that the $d$-spacing of the once-folded form increases in value although the intensity is decreased at the same time as this new phase appears in the WAXD patterns. The crystallization of the butyl branches at the fold surface of the crystalline and amorphous regions would lead to a decrease in the electron density difference between the crystalline and amorphous regions.

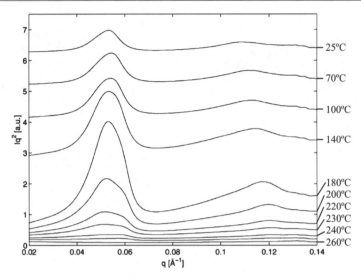

**Fig. 13**  Time-resolved SAXS patterns recorded during cooling at a rate of 4 °C/min at 3.8 kbar show a relatively broad first-order reflection at 118 Å and a second-order reflection at 58 Å. An overall drop in SAXS intensity follows on cooling

It is remarkable that the new phase, once it appears, shows very little expansion or contraction whether caused by temperature or pressure. Again this must be because of the constraints imposed by the folds on the surface, fixing the separation of the branches. Indeed, the orthorhombic reflections can be seen to expand and contract. On heating or release of pressure, when the $d$-spacings of the orthorhombic (110) and the pseudo-hexagonal reflections reach the same value, set by the nearest-neighbor separation determined by a fold, then no further expansion of the lattice is observed, Fig. 14.

## 5.3
### Homogeneous copolymers of ethylene-1-octene

In recent years, ethylene-1-octene copolymers with densities between 870 and 910 kg/m$^3$ have attracted considerable academic and industrial interest. One of the main characteristics of these copolymers is that they are *homogeneous* copolymers, that is, these polymers do not display any differences in comonomer distribution along the chains other than the differences related to statistical fluctuations. Copolymers that do not meet the above definition are considered to be *heterogeneous*, such as LLDPE which shows a superposition of multiple ethylene sequence length distributions (ESLDs). The difference between homogeneous and heterogeneous copolymers lies in the different process of polymerization that result in quite different chain microstructures and final product properties. The homogeneous copolymers are synthesized

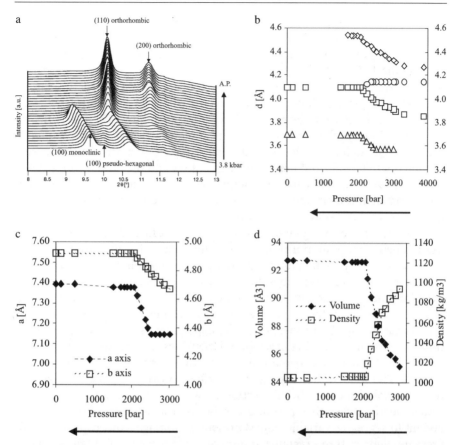

**Fig. 14 (a)** 3D WAXD plot shows that, upon releasing pressure at room temperature, (100) monoclinic and (110) orthorhombic reflections move to lower angles while the (100) pseudo-hexagonal reflections stays at the same position. The latter finally vanishes followed by the disappearance of the monoclinic reflection, resulting in an overall increase in the intensity of the orthorhombic reflections **(b)** The $d$-values for the various crystalline reflections plotted against pressure. Symbols represents: (100) monoclinic ($\diamond$), (100) pseudo-hexagonal (o), (110) orthorhombic ($\square$), (200) orthorhombic ($\triangle$). Upon releasing pressure at room temperature, the $d$-value of the (110) orthorhombic reflection increases from 3.85 Å to 4.08 Å while the $d_{100hex} = 4.16$ Å merges with that of the (110) orthorhombic reflection at $\sim 2.0$ kbar. At atmospheric pressure the $d$-values for the various crystalline reflections attain values similar to those in linear alkanes **(c)** The $a$- and $b$-axis of the orthorhombic unit cell increase and subsequently become constant below 2.0 kbar upon releasing the pressure **(d)** Volume and density of orthorhombic unit cell plotted as a function of pressure at room temperature. The X-ray wavelength used for these experiments is 0.744 Å

with the aid of a single set of reactivity values, corresponding to a single active catalyst site, resulting in a single-peaked ESLD. Whereas synthesis of heterogeneous copolymers involves the presence of two or more active catalyst sites, resulting in an overall distribution of multiple ethylene sequence lengths that causes a multiply-peaked ESLD. Crystallization kinetics and the subsequent morphology are strongly influenced by the specific chain microstructure, specifically the ESLD, which varies with the mole fraction of octene. Especially in the case of ethylene-1-octene copolymers, the chain microstructure largely affects the properties, since the ethylene units will form crystallites, while the octene units are likely to be excluded from the crystallites. Therefore, the exact distribution of the comonomer units along the chain, and consequently the way of stringing together the ethylene sequences determines the crystallizability of that chain and hence the morphology and the ultimate properties. As is evident, the chain microstructure will also influence the chain conformation in the melt and in the solution because the bond angles of the monomers and comonomers will differ.

Ethylene octene copolymers show a very interesting and intriguing crystallization behaviour and morphology upon increasing the comonomer content from zero to very high values, for example, the morphology changes from a lamellar one, organized in spherulites, to a granular-based morphology. Even for homogeneous copolymers there is always a distribution of crystallizable units present in the copolymer chains, ethylene in the present case, and thus variations in the properties of the copolymers such as the crystallization temperature, resulting crystallite dimensions and melting temperature. Clearly, the term homogeneous only applies to the way polymerization is performed. The kind of heterogeneity described above complicates studies of fundamental properties. In this section we will summarize some of our data, which clearly shows that pressure can be used as a thermodynamic component to segregate variable ESLDs present in homogeneous copolymers. By pressure–temperature crystallization cycles it is feasible to obtain a regular organization of branches on the crystal surface and adjacent chain reentry, which may result in the appearance of a new crystalline phase similar to that observed in the case of branched alkanes [51].

In the above section we have shown that it is possible to control the number of entanglements per unit chain in a flexible linear chain polymer such as polyethylene. This could be achieved either from dilute solution, in the melt by high chain mobility along the $c$-axis of the unit cell or by polymer synthesis. The reduction in the number of entanglements, below a critical concentration, promotes drawability of intractable polymers like UHMW-PE having molar mass greater than a million. It is also shown that the drawability of UHMW-PE strongly depends on chain reentry at the interphase. Moreover, on crystallization from the melt, adjacent chain reentry resulting in crystallographic registration of folds at the crystal surface can occur in branched alkanes or homogeneously synthesized ethylene–octene copolymers.

In ethylene-1-octene copolymers, the phase behaviour at elevated pressures is complicated and has been little studied. High pressures and temperatures lead to extended ethylene sequence crystals (EESCs), i.e. the chain sequence can only extend between branching points, if the polymer chain mobility is high enough, rather than extended chain crystals (EECs). By high pressure DSC, Vanden Eynde et al. [53] showed that the melting and crystallization transitions are shifted to higher temperatures due to a shift in the Gibbs free energy of the melt with increasing pressure. The heating rate at elevated pressures had no significant effect on the thermal behaviour except to reveal that the process by which the ethylene sequences extend is very fast. Similarly, increasing the cooling rate leads to broadened DSC peaks as small and imperfect crystallites are formed. For low 1-octene comononer content, three processes could be resolved: the melting of folded chain crystals, followed by the melting of the relatively large EECs/EESCs superimposed on the orthorhombic to hexagonal transition and a high-temperature melting peak of the hexagonal phase. The lack of structural evidence, which the DSC data could not provide, was addressed by in situ high-pressure SAXS and WAXD measurements.

Recently, we investigated the effect of pressure on the crystallization of homogeneous ethylene-1-octene copolymers containing 5.2 mol % and 8.0 mol % 1-octene [54]. Samples that were crystallized at atmospheric pressure from the melt showed a disappearance of the wide-angle diffraction peaks and an increase in the amorphous halo upon increasing pressure to 3.7 kbar at room temperature. However, this amorphization (loss of crystallinity) is more likely to be due to the break up of the large orthorhombic/monoclinic crystallites into small, imperfect crystals at pressures up to 4.0 kbar, as revealed by in situ high-pressure Raman measurements, rather than to be similar to the pressure-induced amorphization reported earlier by one of us for the polymer poly-4-methyl pentene-1 [55–57]. In the latter case, the amorphization at room temperature with increasing pressure is a consequence of an inverse density relationship, i.e. the crystalline density is less than the density of the amorphous region.

In a similar manner, the ethylene–octene copolymer crystallized directly via the orthorhombic phase without the intervention of the anticipated hexagonal phase as would be anticipated in linear polyethylenes at these high pressures and temperatures (at approximately 3.8 kbar and around 200 °C). At $\sim$ 100 °C, see Fig. 15, the $d$ values for (110) and (200) orthorhombic reflections are 4.08 Å and 3.71 Å. When the sample is cooled below 100 °C, a new reflection adjacent to the (110) orthorhombic peak appears at $\sim$ 80 °C. The position of the new reflection is found to be 4.19 Å and so corresponds to a new phase. No change in the intensity of the existing (110) and (200) reflections is observed, however the intensity of the amorphous halo decreases, which suggests that the appearance of the new reflection ($d = 4.19$ Å) is solely due to the crystallization of a noncrystalline component. On cooling further as the new reflection intensifies, the (110) and (200) orthorhombic reflections shift gradually. However, at $\sim$ 50 °C, the (100) monoclinic reflection appears with a concomitant decrease in the intensity of the (110) orthorhombic reflection and a sudden shift in the Bragg $d$ values of the orthorhombic reflections. These results indicate a solid–solid phase transition at 50 °C of a large amount of crystals, from the orthorhombic to monoclinic phase. At room temperature it can be seen that the new reflection is much more intense than the (110) orthorhombic reflection and a corresponding additional new reflection is observed at $d = 3.78$ Å, as shown in Fig. 15.

If one considers the ratio of the two new reflections (4.19 Å/3.78 Å $\approx$ 1.108) and compares their intensities, it can be concluded that this phase resembles an orthorhombic phase (for linear polyethylene, $d_{110ortho}/d_{200ortho} \approx$ 1.111). However, the $d$ values for the new reflections are higher than the conventional $d$ values for the orthorhombic phase and do not match exactly with the known triclinic phase in linear polyethylene. Keeping this in mind and assuming no change along the $c$-axis, the unit cell dimensions for the new orthorhombic phase at 3.8 kbar and room temperature (25 °C), were calculated to be $a = 7.56$ Å, $b = 5.03$ Å and $c = 2.55$ Å. The unit cell volume was found to be approximately 96.97 Å$^3$ and the density 960 kg/m$^3$. This is the first time that this type of orthorhombic phase has been observed in polyethylene. Compared to the conventional orthorhombic phase, the density of the

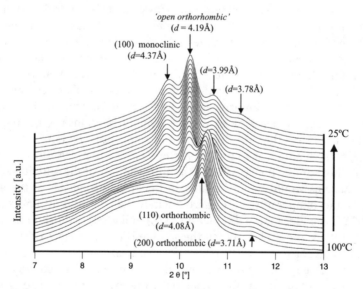

**Fig. 15** Diffraction patterns of ethylene-1-octene copolymer (5.2 mol %) shown from 100 °C to 25 °C while cooling at 10 °C/min recorded during crystallization from melt at $\sim$ 3.8 kbar. The open-orthorhombic phase appears at $\sim$ 80 °C, intensity and position of this reflection remains unchanged. The open-orthorhombic phase is followed by the incoming of the (100) monoclinic reflection concomitant with a shift to higher angles and drop in the intensity of the (110) dense-orthorhombic reflection. The X-ray wavelength used for these experiments is 0.744 Å

new orthorhombic phase is rather low, i.e. the unit cell is more open, for comparison see Table 3. The high intensity of the new open-orthorhombic phase at these high pressures means that the new phase is formed by the crystallization of the majority of the amorphous component with the hexyl branches at the surface of the crystal, i.e. the interphase crystallizes.

The relatively high intensity of the open-orthorhombic phase, at these high pressures, means that this phase cannot be attributed to crystallization of the hexyl branches at the crystal surface alone – unlike in the case of branched alkanes. However, similar to branched alkanes, the fold surface incorporating hexyl branches must be a prominent factor in forming the open-orthorhombic phase. In fact one envisages a major part of the orthorhombic lattice, which resorts to a less dense packing even at these high pressures, as driven by the hexyl branches [58]. In this manner, the stresses due to crystallization of the hexyl branches residing on the fold surface lead to a further compression of the original orthorhombic lattice (termed the *dense orthorhombic phase* at $d_{110} = 3.99$ Å) and a partial conversion of this phase into a monoclinic phase.

On releasing the pressure, the multiple reflections observed in the sample crystallized at elevated pressure and temperature merge into the 110 and 200 reflections of a single orthorhombic phase. If pressure is once again increased at room temperature, Fig. 16, the (110) reflection splits into two distinct $d$ values, $d = 4.19$ Å (open-orthorhombic phase) and $d = 3.99$ Å (dense-orthorhombic phase). The intensity of the dense-orthorhombic reflection ($d = 3.99$ Å) decreases with the appearance of the (100) monoclinic reflection ($d = 4.48$ Å).

From Fig. 16 it can be further confirmed that, with the appearance of the monoclinic phase, the intensity and position of the open-orthorhombic reflection ($d = 4.19$ Å) remains unchanged and does not shift with com-

**Table 3** Comparison of the crystalline lattice parameters, volumes and densities for various polyethylenes

|  | Crystallization Conditions | $a$ Å | $b$ Å | $c$ Å | Volume Å$^3$ | Density kg/m$^3$ |
|---|---|---|---|---|---|---|
| Linear polyethylene | Atmospheric pressure, 25 °C | 7.40 | 4.94 | 2.55 | 93.52 | 996 |
| Branched Polyethylene (ethylene-1-octene 5.2 mol %) | Atmospheric pressure, 25 °C | 7.52 | 4.99 | 2.55 | 95.63 | 974 |
| Branched polyethylene (dense-orthorhombic) | 3.8 kbar 25 °C | 7.16 | 4.81 | 2.55 | 87.82 | 1060 |
| (open-orthorhombic) | 3.8 kbar 25 °C | 7.56 | 5.03 | 2.55 | 96.97 | 960 |

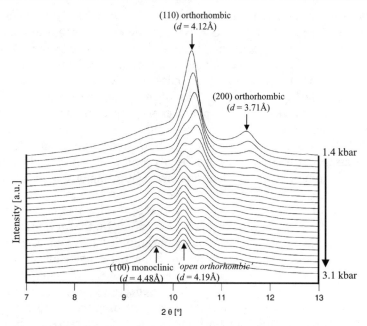

**Fig. 16** Wide-angle X-ray diffraction patterns showing the reappearance of the three crystalline phases namely monoclinic, open-orthorhombic and dense-orthorhombic, during increase in pressure at room temperature

pression. Thus the volume and density of the open-orthorhombic unit cell is invariant with pressure, an unusual feature compared to the changes in volume and density of the dense-orthorhombic phase. These observations strongly suggest that the origin of the open-orthorhombic phase is the crystallization of the fold surface and the interphase, where pressure facilitates the organization process of the fold surface and the interphase to the extent that it could be detected by X-rays diffraction from the bulk. This is possible because pressure facilitates the disentanglement process, phase separation of different molecular weights, reorganization of chains at elevated pressure-temperature and formation of reasonably large crystals along the *ab*-plane in the copolymers and branched alkanes. Thus pressure favours the achievement of well-defined crystallographically registered fold surfaces at the interphase, which on compression crystallizes and gives rise to specific reflections in the WAXD pattern. Rastogi et al. [54] discuss the experimental series above in more detail and from such studies draw the conclusion that pressure facilities the achievement of a well defined crystallographic registered fold surface and the interphase. A shrinkage in the unit cell is observed that suggests a lattice contraction with the concomitant appearance of new reflections – a result in full agreement with recently reported results on lattice contraction with chain unfolding in n-alkanes.

# 6
# Beyond flexible polymers: rigid amorphous fraction

The discussion of the influence of the interphase need not be limited to just linear polyethylenes. Interphases of several nm have been reported in polyesters and poly-hydroxy alkanoates. One major difference between the interphase of a flexible polymer like polyethylene and semi-flexible polymers like PET, PEN and PBT is the absence of regular chain folding in the latter materials. The interphase in these semi-flexible polymers is often defined as the rigid amorphous phase (or rigid amorphous fraction, RAF) existing between the crystalline and amorphous phases. The presence of the interphase is more easily discerned in these semi-flexible polymers containing phenylene groups, such as polyesters.

Similar to polyethylenes the morphology of these polymers is also described as a lamellar stack of crystalline and non-crystalline layers. This so called two-phase model is applied for the interpretation of X-ray diffraction data as well as for heat of fusion or density measurements. However, it is well known that several mechanical properties, as well as the relaxation strength at the glass transition temperature, cannot be described by such a simplistic two-phase approach, as discussed by Gupta [59]. From standard DSC measurements [60], dielectric spectroscopy, shear spectroscopy [61], NMR [62], and other techniques probing molecular dynamics at the glass transition ($\alpha$-relaxation) temperature, the measured relaxation strength is always smaller than expected from the fraction of the non-crystalline phase. The difference in mobility is caused by different conformations of the chains as detected by IR [63] and Raman spectroscopy, or due to spatial confinement because of the neighboring lamellae. To explain the disagreement between the expected values of relaxation strength and the measured values Wunderlich and coworkers [60] proposed a three-phase model. In their approach, the non-crystalline phase is subdivided into one part that does contribute and a second part that does not contribute to the relaxation strength at the glass transition temperature. In addition, Wunderlich and coworkers distinguished between a mobile and a rigid fraction of the polymer. The rigid fraction consists of the crystalline phase and that part of the amorphous phase that does not contribute to the glass transition. This results in a three-phase model consisting of the crystalline (CRF), the rigid amorphous (RAF) and the mobile amorphous (MAF) fractions.

Although the existence of the RAF seems to have been well demonstrated experimentally, the properties of this part is under investigation. One of the most important questions is: what is the chain packing in the RAF and how does it differ from that in the regular MAF? It is also interesting to know whether the chain packing in the RAF is defined by the crystallization conditions (crystallization temperature, nucleation density, pressure etc). Schick et al. [64] demonstrated that the RAF in semi-crystalline PET does not exhibit a separate glass transition temperature, in the entire temperature range up to the melting temperature, $T_m$, while the parameters of sub-$T_g$ relaxation for RAF and MAF are essentially the same. Recently, using temperature-modulated DSC, Schick et al. [65] also showed that similar phenomena take place for several other polymers such as bisphenol-A polycarbonate and poly(3-hydroxybutyrate). No changes in the amount of RAF occurred in the temperature range between crystallization and the glass transition temperatures. Therefore, they suggested that the amorphous chains, constrained between the crystalline lamellae in PET, become effectively vitrified upon crystallization, despite a high temperature, while the remaining amorphous chains located between the lamellae stacks continue to be in the liquid state. Therefore, the crystallization temperature has to be considered as an effective vitrification temperature for the RAF. Devitrification of the RAF should then occur upon melting of the crystalline lamellae, consisting of the lamellae stacks.

If this hypothesis is right, the specific volumes that characterize the RAF and MAF have to be essentially different below the crystallization temperature. Figure 17 exhibits a sketch to illustrate this point. This sketch basically shows a hypothetical thermal-expansion behavior associated with the RAF and MAF for PET, crystallized at some arbitrary crystallization temperature, $T_c$. Above $T_c$, in the equilibrium melt, only one phase occurs, i.e. the specific volumes for the RAF and MAF are the same. If vitrification of the RAF occurs at $T_c$, the slope of specific volume versus temperature for this fraction should change at $T_c$, and become characteristic of the glassy state in the temperature interval below $T_c$. In the same manner for the MAF, the slope of specific volume versus temperature, below $T_c$, should continue to be the same as for the equilibrium melt and change only at the real $T_g$. Therefore, if room temperature (25 °C) is considered as the reference, the specific volume for the RAF at 25 °C must be larger than that for the MAF. The same reasoning would lead to the anticipation that the specific volume of the RAF will be a direct function of $T_c$.

Lin et al. [66] have exploited this variation in specific volume of the RAF to control the barrier properties of polyester films. An attempt to correlate the mechanical deformation of PET with the amount of RAF present in these films has been made recently. Moreover, the observations have been that a sample with a larger amount of RAF, on uniaxial compression, shows considerable loss in crystallinity compared to a sample having a lower amount of RAF. These findings have been reported in a recent publication [67].

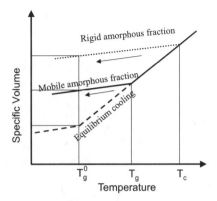

**Fig. 17** Representation of the volume–temperature relationship for rigid amorphous and mobile amorphous phases of PET

# 7
# Conclusions

In this chapter we have discussed the morphological aspects of the interphase that exists between the ordered three-dimensional crystalline phase and the randomly structured amorphous phase. From a series of experimental observations it is evident that the structure of the interphase strongly depends on the crystallization conditions, for example, a chain from a crystal can be reentrant to the crystal of origin, with or without forming entanglement(s). The amount of entanglements present on the crystal surface can be controlled either by crystallization from dilute solution, by enhancing chain mobility along the $c$-axis of the crystal through the hexagonal phase or during synthesis. Chain sliding diffusion along the $c$-axis, which causes lamellar thickening, shows a strong dependence on chain topology. Ultimately topological constraints present on the crystal surface control the macroscopic deformation behaviour of polymer films.

Non-adjacent re-entrant chains build up steric stresses, which are metastable and tend to decrease in number when given sufficient time and chain mobility for reorganization. However, when the chain mobility within the lattice is inhibited with the introduction of side branches, the chains tend to adopt adjacent re-entry at the surface during crystal growth. This results in minimization of steric stresses at the crystal basal plane, thus favoring extensive lateral growth most likely by an adjacent reentran chain-folding process. Thus side branches of the homogeneous copolymers and branched alkanes tend to order on the $ab$-plane of the crystal lattice and will crystallize on compression. Simultaneous to the crystallization of branches, a contraction in the parent crystal lattice occurs. These observations clearly show that extended chain crystals need not be the thermodynamically stable state; minimization

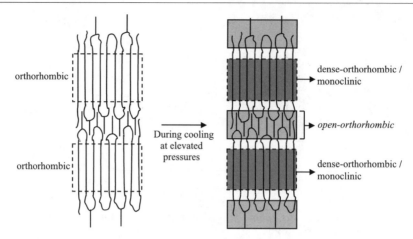

**Fig. 18** Schematic two-dimensional presentation for crystallization of the interphase. With crystallization of the interphase in the open-orthorhombic phase, contraction in the orthorhombic crystalline lattice occurs. This results in the formation of dense ortho-rhombic or monoclinic phase (see Figs. 15 and 16). In branched alkanes, because of the relatively sharp folds, the amount of chains contributing to the interphase is considerably less, which corresponds to the low intensity of the reflection assigned to the pseudo-hexagonal phase (see Figs. 12 and 14), while the phenomenon of concentration of the lattice with crystallization of the interphase is the same as in copolymers

in surface free energy can also be achieved by having an interphase with adjacent re-entrant chains. Figure 18 is a schematic drawing showing such a possibility. The intensity of the crystallizable interphase in the open orthorhombic phase strongly depends on the crystallizable material at the crystal surface and the interphase. In our studies, reported in this chapter, the content of the crystallizable interphase would be considerably higher in ethylene–octene copolymers compared to adjacent re-entrant folds in branched alkanes.

The implications of adjacent re-entry is realised in the drawability of the solution crystallized film of UHMWPE and the strain hardening behaviour of branched ethylene-octene copolymers, for example.

**Acknowledgements** The authors wish to thank Dr. Ankur Rastogi (Dow Chemicals), Dr. Vincent Mathot (DSM Research) for their contribution in development of the work. One of the authors (S. R.) wishes to thank the Max Planck Society and The Dutch Polymer Institute for the financial support. Experimental support provided by the Materials Science beamline ID11 and Ultra Small Angle beamline ID2 (ESRF, Grenoble) is gratefully acknowledged.

# References

1. Flory PJ (1953) Principles of Polymer Chemistry. Cornell University Press, Ithaca, New York
2. Mandelkern L (1983) An Introduction to Macromolecules. Springer-Verlag, New York
3. Mandelkern L (1983) Crystallization in Polymers. McGraw-Hill, New York
4. Mandelkern L (1992) Chemtracts (Macromol Chem) 3:347
5. Baker AME, Windle AH (2002) Polymer 42:667
6. Gautam S, Balijepalli S, Rutledge GC (2000) Macromolecules 33:9136
7. Bassett DC, Hodge AM (1981) Proc R Soc London A377:25; ibid (1981) A377:39; ibid (1981) A377:61
8. Khoury F (1979) Faraday Discuss Chem Soc 68:404
9. Frank FC (1979) Faraday Discuss Chem Soc 68:7
10. Yoon DY, Flory PJ (1984) Macromolecules 17:868; ibid (1984) 17:862
11. Kumar SK, Yoon DY (1989) Macromolecules 22:3458
12. Mandelkern L (1990) Acc Chem Res 23:380
13. Bassett DC, Hodge AM (1981) Proc R Soc London A377:25
14. Keith HD, Padden FJ (1996) Macromolecules 24:7776
15. Toda A, Okamura M, Hikosaka M, Nakagawa Y (2000) Polymer 44:6135
16. Balijepalli S, Rutledge GC (1998) J Chem Phys 109:6523
17. Smith P, Lemstra PJ, Booij HC (1982) J Polym Sci B Polym Phys 20:2229
18. Lemstra PJ, Bastiaansen CWM, Rastogi S (2000) In: Salem DR (ed) Structure formation in polymeric fibers. Hanser, p. 185
19. Ward IM (1988) Developments in oriented polymers, 2nd ed. Elsevier, New York
20. Bassett DC (1976) Polymer 17:460
21. Wunderlich B, Grebowicz J (1984) Adv Polym Sci 60/61:1
22. Hikosaka M, Rastogi S, Keller A, Kawabata H (1992) J Macromol Sci Phys Ed B31:87
23. Rastogi S, Hikosaka M, Kawabata H, Keller A (1991) Macromolecules 24:6384
24. Hikosaka M, Tsujima K, Rastogi S, Keller A (1992) Polymer 33:2502
25. Maxwell AS, Unwin AP, Ward IM (1996) Polymer 37:3293
26. Smith P, Chanzy HD, Rotzinger BP (1985) Polym Comm 26:258
27. Rastogi S, Kurelec L, Lemstra PJ (1998) Macromolecules 22:5022
28. Rastogi S, Kurelec L, Lippits D, Cuijpers J, Wimmer M, Lemstra PJ (2005) Biomacromolecules 6:942
29. Ostwald W (1897) Z Physik Chem 22:286
30. Uehara H, Yamanobe T, Komoto T (2000) Macromolecules 33:4861
31. Rastogi S, Spoelstra AB, Goossens JGP, Lemstra PJ (1997) Macromolecules 30:7880
32. Xue YQ, Tervoort TA, Rastogi S, Lemstra PJ (2000) Macromolecules 33:7084
33. Ungar G, Zeng X (2001) Chem Rev 101:4157
34. Terry AE, Phillips TL, Hobbs JK (2003) Macromolecules 36:3240
35. Schmidt-Rohr K, Spiess HW (1994) in: Multidimensional Solid-State NMR and Polymers. Academic, London, p. 478
36. Brooke GM, Burnett S, Mohammed S, Proctor D, Whiting MC (1996) J Chem Soc Perkin Trans 1:1635
37. Wunderlich B (1980) Macromolecular Physics, Vol. 3: Crystal Melting. Academic, New York
38. Kortleve G, Tuijnman CA, Vonk CG (1972) J Polym Sci B Polym Phys 10:123
39. Hosoda S, Nomura H, Gotoh Y, Kihara H (1990) Polymer 31:1999
40. Vonk CG, Reynaers H (1990) Polymer Commun 31:190
41. Zachmann HG (1967) Kolloid-Z u Z Polymere 216–217:180

42. Vonk CG (1986) J Polym Sci Polym Lett 24:305
43. Ungar G, Stejny J, Keller A, Bidd I, Whiting MC (1985) Science 229:386
44. Ungar G, Keller A (1986) Polymer 27:1835
45. Organ SJ, Keller A, Hikosaka M, Ungar G (1996) Polymer 37:2517
46. Zeng X, Ungar G (1998) Polymer 39:4523
47. Ungar G, Zeng X, Brooke GM, Mohammed S (1998) Macromolecules 31:1875
48. Zeng X, Ungar G (1999) Macromolecules 32:3543
49. Spells SJ, Zeng X, Ungar G (2000) Polymer 41:8775
50. Hikosaka M, Seto T (1982) Jpn J Appl Phys 21:L332
51. Rastogi A, Hobbs JK, Rastogi S (2002) Macromolecules 35:5861
52. Hay IL, Keller A (1970) J Polym Sci C 30:289
53. Vanden Eynde S, Mathot VBF, Hoehne GWH, Schawe JWK, Reynaers H (2000) Polymer 41:3411
54. Rastogi A (2002) PhD Thesis, Eindhoven University of Technology; Rastogi A, Terry AE, Mathot VBF, Rastogi S (2005) Macromolecules 38:4744
55. Rastogi S, Newman M, Keller A (1991) Nature 353:55
56. Rastogi S, Newman M, Keller A (1993) J Polym Sci B Polym Phys 31:125
57. Rastogi S, Hoehne GWH, Keller A (1999) Macromolecules 32:8897
58. It is to be noted that the reflection assigned to the "new phase" in butyl branched alkanes is relatively weak compared to the reflections observed for the "new phase" in ethylene-1-octene copolymer (5.2 mol %). As explained in this chapter, we attribute the "new phase" to the crystallization of transient layer (butyl branches and fold surface). Considering the anticipated tight folds for butyl branched alkanes, the amount of crystallizable entities in the branched alkanes would be much less than in the ethylene-1-octene copolymers where the loose folds are expected. We would like to mention that, considering the $d$-value and intensity of the pseudo-hexagonal phase in branched alkanes, this reflection may be referred to as open-orthorhombic phase.
59. Gupta VB (2002) J Appl Polym Sci 83:586
60. Suzuki H, Grebowicz J, Wunderlich B (1985) Makromol Chem 186:1109
61. Huo P, Cebe P (1992) Macromolecules 25:902
62. Gabriels W, Gaur HA, Feyen FC, Veeman WS (1994) Macromolecules 27:5811
63. Cole KC, Aiji A, Pellerin E (2002) Macromolecules 32:770
64. Schick C, Dobbertin J, Potter M, Dehne H, Hensel A, Wurm A, Ghoneim AN, Weyer S (1997) Therm Anal 49:499
65. Schick C, Wurm A, Mohamed A (2001) Colloid Polym Sci 279:800
66. Lin J, Shenogin S, Nazarenko S (2002) Polymer 43:4733
67. Rastogi R, Vellinga WP, Rastogi S, Schick C, Meijer HEH (2004) J Polym Sci B Polym Phys 42:2092

# Author Index Volumes 101–181

Author Index Volumes 1–100 see Volume 100

# Subject Index

Adv Polym Sci (2005) 180: 221
DOI 10.1007/017
© Springer-Verlag Berlin Heidelberg 2005
Published online: 29 June 2005

# Erratum to
# Effect of Molecular Weight and Melt Time and Temperature on the Morphology of Poly(tetrafluorethylene)

P. H. Geil[1] (✉) · J. Yang[1,2] · R. A. Williams[1] · K. L. Petersen[1] · T.-C. Long[3] · P. Xu[3]

[1]Department Of Materials Science And Engineering, University Of Illinois, Urbana, IL 61801, USA
*geil@uiuc.edu, junyan@engin.umich.edu, rawilli1@uiuc.edu, klpeter1@uiuc.edu*

[2]Department Of Materials Science And Engineering, University Of Michigan, Ann Arbor, MI 48109, USA
*junyan@engin.umich.edu*

[3]W. L. Gore and Assocs., Ltd., Newark, DE USA
*tlong@wlgore.com, pxu@wlgore.com*

Unfortunately, instead of "Fig. 19" the publisher printed "Fig. ??" on pages 109 and 111 in this volume.

Printing: Krips bv, Meppel
Binding: Stürtz, Würzburg